Illustrating Science

ILLUSTRATING

Scientific Illustration Committee

Paul J. Anderson, Chairman
Arly Allen
Margaret Broadbent
Bradley Hundley
Christopher Johnson
Frances H. Porcher
Laurel V. Schaubert
Vivian B. Rhoads
G. William Teare, Jr.

SCIENCE:

STANDARDS FOR PUBLICATION

Prepared by the

SCIENTIFIC ILLUSTRATION COMMITTEE
of the Council of Biology Editors

Published by the

COUNCIL OF BIOLOGY EDITORS
Bethesda, Maryland

COUNCIL OF BIOLOGY EDITORS · INC.
BETHESDA · MD 20814

PRINTED IN THE UNITED STATES OF AMERICA

LIBRARY OF CONGRESS CATALOGING IN PUBLICATION DATA

CBE Scientific Illustration Committee
Illustrating Science: Standards for Publication

Bibliography: p.
Includes index

1. Scientific illustration—editing. 2. Illustration—preparation for printing. 3. Line art—printing. 4. Halftone—printing. 5. Color illustrations—printing. 6. Computer graphics. 7. Ethical and legal considerations. 8. Glossary.
I. Title. II. Council of Biology Editors.

Q222.C68 1988
502'.2—dc19

ISBN: 0-914340-05-0 88-9557

CONTENTS

PREFACE

The purpose of scientific publication is to communicate knowledge gained through scientific research. As technology advances, the expectations of prompt, accurate, and cost-effective delivery of information are being challenged. The traditional forms of published information—print on paper—remain; but electronic media are expanding the options and new forms of "publication" are emerging.

While technology is expanding the ways in which the results of scientific research are communicated, the human elements involved in the process continue to play key roles. This process will continue to follow a traditional pattern: authors will investigate, discover and report, illustrators and photographers will depict and interpret, reviewers will criticize and recommend, editors will judge and select, publishers will organize and manage, and printers will produce the finished product to be delivered in print or electronic form.

The contributors to this book have broad experience in the preparation and evaluation of scientific illustrations. Each chapter is the result of lengthy work sessions and discussions which revealed aspects of preparation and production of illustrations that no single individual had experienced in totality. We learned that criteria do indeed exist—the author's, the illustrator's, the editor's, the photographer's, the publisher's, the printer's, and the reader's—but they exist with differing shades of emphasis. It became evident that although criteria were clearly followed, individual perspectives often failed to encompass the complete view.

The purpose of this book, then, is to develop specific standards and guidelines for publication of illustrated scientific material. Our goal is not to teach the fine points of executing an illustration; rather it is to guide the judgment of an illustration's worth. We further wish to help each professional involved in preparing scientific illustrations to understand why and how these materials are best produced. We have tried to provide relevant information on the current state of the art. We have sought to set standards for creating and printing illustrations that will achieve the finest possible results with the greatest possible effectiveness. We hope that this book will be a practical and useful resource for the reader who must publish illustrated scientific information.

The project which led to the preparation of this book began with the appointment of an Ad Hoc Committee for Scientific Illustration in 1980 by Erwin Neter, then President of the Council of Biology Editors. The Ad Hoc Committee consisted of Paul J. Anderson (chairman), Arly Allen, Robert Demarest, Bradley Hundley, and Aaron J. Ladman. The

Committee was charged to explore the need for a "style manual" on scientific illustration.

Personal contact was established with the leadership of organizations that represent different areas of specialization in scientific publication. The organizations included the American Medical Publisher's Association, American Medical Writer's Association, Association of Medical Illustrators, Biological Photographer's Association, Guild of Natural Science Illustrators, and the Society for Scholarly Publishing. All responded favorably and encouraged the preparation of a CBE publication on scientific illustration directed especially to the needs of authors and editors.

The membership of the Council of Biology Editors was canvassed by questionnaire to elicit members' opinions on the need for a publication on this topic and their preferences for format and subject matter. Most of the respondents were involved in editorial management of journal publications. Their needs were concerned with production standards for line art, graphs, halftones, and computer graphics. Major concerns were expressed also for quality criteria, methods of proofing, color standards, and instructions for authors.

Finally, a survey of published instructions to authors on the preparation of scientific illustrations for submission to journals was conducted. The guidelines of 465 clinical and basic science journals were reviewed. Fewer than 10% were judged sufficiently detailed to serve as possible models for standardized instructions.

The need for this publication was clear. The recommendations of the Ad Hoc Committee formed the basis for planning, and the present Committee on Scientific Illustration was mandated to implement the project.

Although each member of the Scientific Illustration Committee wrote major sections of the book, others participated in equally significant ways as contributors, consultants, and reviewers.

Among contributors, Margaret Foti and A. Jean MacGregor, both original members of the Committee, deserve special recognition. Ms. Foti, while serving as the Program Chairman for the Council of Biology Editors in 1980, arranged the Council's first symposium and workshop on scientific illustration (most of the Committee members were participants in that program) and promoted the idea of a publication on this subject. Ms. MacGregor's guidance on the preparation of charts and graphs during the planning stages of the project was critical.

Mimi Zeiger prepared the section on graphs (Chapter IV) and supervised the design of each example. Alan Baker and Margaret Kurzius assisted in the preparation of the section on maps (Chapter IV). Graphs for Chapter IV were prepared by Margaret Kurzius and Bobbi Angell. Maps were prepared by Margaret Kurzius.

Valuable contributions on preparation of artwork (Chapter III) were made by Ernest Beck, Robert Demarest, Neil O. Hardy, Elaine R. S.

Hodges, Denis Lee, Nelva Richardson, and Patricia Weigand. Contributors to Chapter IX included Neil O. Hardy, Elaine R. S. Hodges, and Nelva Richardson.

The chapter on computer graphics (Chapter V) was prepared by Robert W. Haley, Director of the Division of Epidemiology at Southwestern Medical School in Dallas, Texas.

Assistance and guidance in the preparation of the chapter on camera ready copy (Chapter VI) were provided by Norman Och. Contributors to the chapter were Karl F. Heumann and Edith C. Wolff.

The sections on photography and halftone production (Chapter VII) contain contributions by Melvin M. Figley, Thomas P. Hurtgen, and Marie Leonard.

J. L. Juergens, Walter G. Peter, and William B. Propert provided assistance in the chapter on color (Chapter VIII).

Consultants and reviewers generously shared their expertise with the Committee, and their counsel shaped many revisions of the text. The Committee is deeply grateful to:

Mary Challinor (American Assn. for the Advancement of Science, Washington, DC)
Nancy Clark (Editorial Consultant, Sausalito, CA)
William S. Cleveland (Bell Laboratories, Murray Hill, NJ)
Emily Craig (Hughston Clinic, Columbus, GA)
Tad Crawford (Counsel, Graphic Artists Guild, New York, NY)
William Conkling (French/Bray, Inc., Glen Burnie, MD)
Norman L. Fisher (E. I. DuPont de Nemours & Co., Wilmington, DE)
Linda Golder (Woods Hole Marine Biology Laboratory, Woods Hole, MA)
Robert Golder (Woods Hole Marine Biology Laboratory, Woods Hole, MA)
Gerald Hodge (University of Michigan, Ann Arbor, MI)
Elaine R. S. Hodges (Smithsonian Institution, Washington, DC)
Edward Huth (Annals of Internal Medicine, Philadelphia, PA)
Lawrence B. Isham (Smithsonian Institution, Washington, DC)
Donald Luce (Bell Museum of Natural History, Minneapolis, MN)
Maeve O'Connor (CIBA Foundation, London, UK)
Robert V. Ormes (American Assn. for the Advancement of Science, Washington, DC)
Howard Radzyner (Radzyner, Inc., New York, NY)
James Ransom (Lange Medical Publications, Los Altos, CA)
Will E. Renner (University of California, Davis, CA)
Carol A. Risher (Association of American Publishers, Washington, DC)

Robert F. Tipton (Science Press, Ephrata, PA)
Roger B. Williams (University of Kansas, Lawrence, KS)

Special thanks are due to several organizations that provided material and intellectual support to the project. Outstanding contributions were made by Allen Press (Lawrence, Kansas), Biomed Arts Associates (San Francisco, California), Capital City Press (Montpelier, Vermont), The William Byrd Press (Richmond, Virginia), Waverly Press (Baltimore, Maryland), and the Printing Industries of America (Arlington, Virginia). Facilities for conducting meetings of the Committee were generously provided by Rockefeller University Press.

The resources and staff of the editorial office of the Journal of Histochemistry and Cytochemistry were also utilized extensively throughout the management of the project. Sharon Rule served as copyeditor for the manuscript. The index was prepared by Phyllis Manner.

Finally, the collaboration of other organizations with professional objectives similar to those of the Council of Biology Editors must be gratefully acknowledged. Shared concerns about the standards of quality for scientific illustration made such liaisons especially important for implementation of the project. The Scientific Illustration Committee and the Council of Biology Editors are especially indebted to the following organizations:

THE ASSOCIATION OF MEDICAL ILLUSTRATORS
THE GUILD OF NATURAL SCIENCE ILLUSTRATORS
THE BIOLOGICAL PHOTOGRAPHIC ASSOCIATION
THE GRAPHIC ARTISTS GUILD

The continued collaboration of these organizations with the Council of Biology Editors will help us correct present deficiencies in future editions of this book.

Scientific Illustration Committee
Paul J. Anderson, *Chairman*
Arly Allen
Margaret Broadbent
Bradley Hundley
Christopher Johnson
Frances Porcher
Vivian B. Rhoads
Laurel V. Schaubert
G. William Teare, Jr.

August 21, 1987

Illustrating Science

CHAPTER I
PREPRESS: GETTING AN ILLUSTRATION READY TO PRINT

This chapter describes how an illustration is developed, from the scientist's initial concept, through its creation, to being ready to print (Fig. 1–1). It gives a brief overview of the publishing operation for the twin purposes of aiding scientists, illustrators, and editors in the preparation of illustrations and of helping them understand the creative, review, and editorial processes through which illustrations pass.

What is an Illustration?

There are two basic kinds of illustrations commonly used in scientific literature: line art and halftone illustrations. Both kinds can be printed in black and white or in color. Black and white illustrations are the most common in scientific work. Black and white line art usually includes charts (1 variable), graphs (2 variables), maps, ink drawings, and anything that is meant to be photographed as black on white. Black and white illustrations include drawings, paintings, or photographs that show gradations of shading. Line art reproduces as black on white; tone illustrations display a range of tones from light to dark. These two kinds of illustration require different methods of production and reproduction. The choice of which process to use is determined by the content of the illustration, the publisher's constraints, and budgetary considerations. Full-color (process) illustrations, or the addition of one or two colors to line copy, will add to the cost of printing but are sometimes necessary to depict the subject accurately and effectively.

The Importance of Communication

Every illustration in a scientific article is handled by many professionals, each in a specific field and each with special concerns and ways of expressing those concerns. Therefore, of all the suggestions that this book offers, there is one that should be considered the watchword: COMMUNICATE.

The process of communication begins with the author and involves the illustrator, the reviewer, the editor, the publisher, the printer, and

FIG. 1–1: *Preparing scientific illustrations.*

1.
Author consults with artist and/or photographer about desired illustrations.

2.
Drawings and/or photographs are produced.

3.
Author sends copies of drawings, photographs, and manuscript to editor. Editor in turn sends copies to reviewers.

4.
When manuscript is accepted, editor and production manager discuss with author any problems with size or quality of illustrations. Those that are unacceptable are returned to author. Author asks original artist and/or photographer to make requested changes.

5.
When illustrations are acceptable, they are sent to the production department.

6.
The production department identifies and sizes illustrations and sends them to the printer.

7.
Printer examines and logs in the illustrations and sends them to the camera.

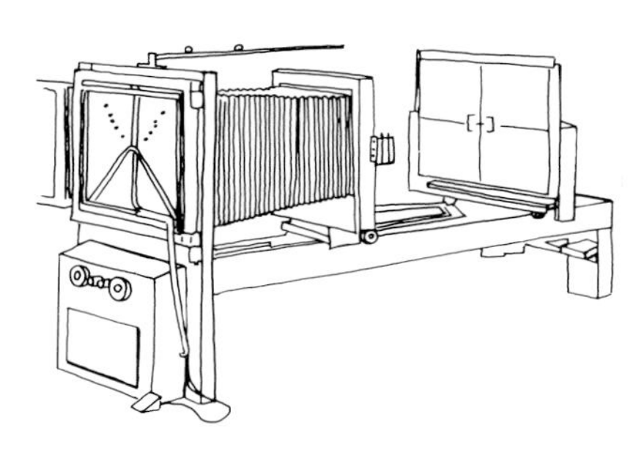

8.
a. Drawings are shot and proofs made.
b. Photographs are shot and proofs made. Original negatives are saved for platemaking.
c. Color separations and proofs are made from color artwork.

9.
One proof is sent for pasteup. Others are sent with original illustrations to author and editor.

10.
Author, editor and artist or photographer check proofs against original illustrations to be sure that clarity, color, contrast, and quality have been maintained.

11.
If changes have occurred during the developing process the original illustrations are returned to production for corrections.

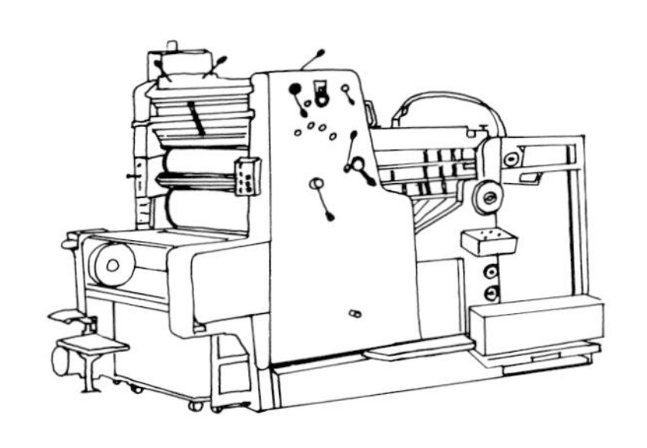

12.
If no changes have occurred, illustrations are ready to be printed.

13.
Book or journal is printed!

finally the reader. At each stage the illustration must be precisely understood. The task of creating this understanding and of achieving the communication intended by the illustration is the focus of this book. In the succeeding chapters we will focus on specific types of illustrations and the methods of presenting them. In this chapter we wish to provide an overview of the process through which an illustration passes from creator to consumer.

Terminology

Because communication within the related fields of editing, publishing, and printing is complicated by the various "languages" of each discipline, a glossary at the back of this book defines words commonly used in illustrating, photographing, printing, and editing. Many of the terms have been with us for centuries, but new words are constantly being introduced as methods and technologies change. The best way to ensure effective communication of needs and ideas is to discuss methods with those in other fields. This will invariably help to improve the level of mutual understanding in the complex task of getting scientific illustrations into print.

The Author as Creator

Illustrations usually originate as the concept of the author, whose first concern is to decide whether or not an illustration is needed. Once the need for an illustration is established, the objective is to present all data clearly and concisely in a manner that will clarify the author's ideas and conclusions. The author may wish to leave the actual execution of the illustration to a professional artist or photographer, although on occasion authors themselves may create the necessary artwork or photography. How successfully this is accomplished often depends on two things. The first is the author's ability to approach the data from the viewpoint of the reader. To be useful, an illustration must convey information more efficiently than words can do. If words would do a better job, it is probably best to omit the illustration.

The second is how well the author is able to execute the design of the illustration. A good concept is often ruined by poor artwork.

In many instances, the particular research area dictates the form of the illustration. However, even when the author has decided on the general form (graph, line art, or tone illustration), the best way to go about producing it may not be clear. In such cases, the author should seek professional advice in organizing data from a statistician or technical editor, or, in executing the illustration, from an illustrator, com-

FIG. I–2: *Illustration problems and their solutions.*

	Problems		Solutions
1	The illustrations are too large to be handled efficiently.	→	Line illustrations should be no more than 200% of reproduction size. Photographs or tone art should be submitted 100% of reproduction size.
2	Illustrations are not labelled or clearly identified.	→	Place a label on the front of the figure but outside the reproduction area. Identify the author, the figure number, the publication (if known) and the top of the figure.
3	Illustrations are borrowed from other publications without permission.	→	Most editors expect all illustrations to be original. But should borrowing be necesssary, written permission must accompany the artwork when it is submitted.
4	The quality is bad. The photographs are blurred; the line illustrations are poorly drawn.	→	Submit only original photos or artwork for reproduction. Have both prepared to professional standards.
5	Lettering is broken, weak or illegible.	→	Have all lettering done in a professional manner. Use a sans serif face like Helvetica. Have letters approximate 8 point type when figure is reduced.
6	The illustration contains numerous special symbols which are not identified within the illustration itself.	→	When symbols are used in an illustration, they should be identified in a key placed within the illustration itself. Printers may not be able to match symbols created by illustrations.
7	A color illustration is submitted to be printed in black and white.	→	If black and white reproductions are required, submit black and white illustrations. If color reproductions are wanted, submit color illustrations.

puter operator, or photographer. Even though the author has an idea clearly in mind, consultation with a professional artist or photographer often leads to a clearer and more comprehensive illustration than the initial concept.

The Illustrator

The illustrator may be a graphic artist or photographer who is primarily concerned with capturing the author's idea visually. To create an illustration that communicates information clearly, the illustrator or photographer must fully understand the author's goal and the content of the text. Occasionally this means that the illustrator must be involved with the project almost from its inception. In some cases, the illustration "medium" may be an electrocardiographic tracing, the tracing of a sound pattern, or other machine-generated print. In other cases, however, the illustrator or the photographer can create original art that greatly improves the communication of data. The key to success here is that the author must take the time to provide the illustrator or the photographer with detailed information about the subject and about what the illustration is intended to do.

When new data are presented, the illustration should be kept simple. The essential elements must be defined and "decoration" weeded out; too much detail is as unsatisfactory as too little. The goal is to make the complex seem simple. The professional illustrator or photographer can do this with an illustration, but it requires advance planning. The author should confer with the illustrator or photographer well ahead of any deadlines to allow adequate time for the design and development of the illustration. Written specifications, essential at the production and printing stages, are equally useful at the planning stage (see Chapter III). If written instructions would be difficult to follow, the author should provide examples of illustrations taken from other texts, if these are available.

Whenever possible, the author should make the illustrations and text conform to the style of the publication to which the material will be submitted. In the case of books, format is seldom predetermined as it is with journals. In either type of publication, when there are questions about format, the author should consult the editor or production manager *before* having the illustrations prepared. The more care the author and the illustrator take in complying with the publication's requirements, the more successful and the less costly the final printed illustration is likely to be.

The Reviewer

An illustration intended for publication in a scientific journal or book is an integral part of the manuscript. It will therefore be evaluated by editors and reviewers as a part of a whole, not as a separate entity. Reviewers play a crucial role in determining what will and what will not be published. They bring a fresh and critical eye to the data and their manner of presentation. A good reviewer must understand that illustrations (often received in the form of photocopies) are a basic part of the research and must be evaluated as carefully as the text. The author, in turn, should take care to provide the reviewer with artwork that is clearly reproduced and as close as is feasible to its final form.

One of the reviewer's tasks is to determine whether all of the illustrations are really necessary. A distinction should be made between illustrations that are essential and those that are supportive but nonessential. Let Occam's razor serve as a guideline: more are superfluous when less will serve.

A second task for the reviewer is to judge whether the illustrations present accurate and clear information. Do they make their points effectively? Is the information consistent with the text? Do the illustrations overstate the case or mislead the reader? Illustrations that do not support the conclusions of the text should be omitted.

Design

Reviewers and editors are often uncomfortable making judgments about design, for example, the arrangement of the elements of a graph, a drawing, or a photograph. Yet the purpose of the illustration is to communicate information that lies within the competence of both the reviewer and the editor. For good or ill, design affects this transfer of information and must be evaluated as closely as is the text. There are three main considerations for evaluating an illustration: (1) the effectiveness of the internal design; (2) the consistency of design from one illustration to another; and (3) the cost of printing.

The effectiveness of internal design

An effective illustration should be approachable on two different levels. The first level is the immediate: the point of the illustration must be apparent at once. For this reason the illustration should not be overburdened with information, but it should be selective in its presentation. The reader should not have to work to understand it. The second level of an illustration is the substantive: the reader should be able to study the illustration in depth and to test its conclusions by reference to other points within the illustration or in the text itself. Although simplicity of design is desirable, omitting critical detail for the sake of simplicity may result in failure to present the conclusions convincingly. Accuracy and intelligibility are therefore functions of design. Illustrator, photographer, and author must recognize these two levels of information and present both within the illustration.

Tonal contrast within an illustration is another essential design feature that affects the illustration's usefulness. If a photograph lacks suitable contrast, detail will be lost in the printed version. If both subject and background are dark, it may be difficult to identify the subject; a similar problem exists when subject and background are both light. The primary responsibility for achieving proper contrast is the photographer's. In examining the photographs, the editor and reviewer, who are frequently already familiar with the material under review and know what they are looking for, should keep in mind that the reader must be able to comprehend the author's message with little or no prior knowledge. The image, by proper use of contrast, must clearly communicate the facts.

The same points are true of a drawing. The design must be structured so that the reader can easily grasp the significant features. If shading is used that either "drops out" or "plugs up" in the reproduction process, detail is lost and the illustration becomes confusing or meaningless. The more effort the reader must expend, the less likely it is that the information will be useful or used.

Consistency of design

Consistency in illustrative material is often difficult to achieve. At the same time it is one of the most important design features. It is easier to understand a series of illustrations when all have the same size, scale, and proportions. Inconsistency in size from illustration to illustration is visually annoying and may actually be misleading, since difference in size suggests difference in importance. Thus, a large illustration coupled with a small illustration may suggest that the information in the larger is more important than that in the smaller. Alternatively, the reader may misinterpret the small illustration as a subset of the larger one. Consistency of design and format avoids these problems and helps both the author and the reader. In some instances, and certainly when the difference is intended to convey relative importance, varied illustration size may be the best solution (for example, an enlargement of the pertinent portion of a subject).

The problems of achieving consistency in size and format of illustrations are complicated by the fact that illustrations must often be reduced to fit within the printed page. If illustrator and photographer present the author with illustrations whose range of sizes is too great, even though they are consistent with each other, the process of reduction for printing may be uneven, resulting in inconsistent sizing. The reviewer and the editor should take this problem into consideration when evaluating a manuscript. When possible, the author and the artist or photographer should refer to the Guidelines to Authors (see Chapter x) in advance so that materials can be properly sized at the outset. Without such guidelines, authors cannot be expected to submit properly prepared artwork.

Cost

It follows from the preceding discussion that if illustrations are concise, correct, complete, with suitable contrast, consistent (with each other), and in compliance with the instructions of the publication, the cost of production will be less. If illustrations are prepared with the same care with which the text should be prepared, production costs will be kept down. If they are an afterthought, production costs will probably increase. This increased cost will be borne by the author or reader of scientific publications, and by the illustrator or photographer if their work is rejected as unsatisfactory. It will be borne by the editor and reviewer in the time they must spend attempting to correct errors. In short, all lose and no one gains when the basic rules of clear and correct illustration are not followed.

The Editor

With the reviewer's critique in hand, the editor can make decisions concerning revision of text and illustrations. The editor must determine how to fit the illustrations within the format required, for the most efficient and useful presentation. If any illustrations do not meet the production department's requirements, they must be returned with instructions for alterations. This requires that the editor have a good working knowledge of how the production and printing operations function, so that delays can be avoided and costs reduced by rejecting illustrations that cannot be suitably reproduced.

The Publisher and Printer

The publisher is responsible for the planning and publication of the material; the printer is responsible for the technical details of photographing and reproducing illustrations. Their work supplements the work of the author, the author's editor, the scientific editor, and the production editor. As a paper progresses from manuscript to print, the author, editor, reviewer, production staff, and the printer's technically trained personnel are all involved with specific aspects of producing text and illustrations. After a manuscript and its illustrations have been accepted for publication, they are passed to a copy editor.

The copy editor's job is to facilitate the production of the text. However, the copy editor may also handle the illustrations, although in large companies this responsibility is sometimes turned over to an art department. In either case, the figures are again examined for quality, corrected professionally if necessary, and marked (sized) for the appropriate production size. It should be noted that under the terms of the new copyright law, if a drawing is signed or copyrighted, only the original artist has the legal right to make changes in it.

It is particularly important to point out that changes made by the copy editor in the text may affect the illustrations, while changes in the illustrations may affect the meaning of the text. Thus it is important that any alterations in copy be carefully considered in conjunction with the illustrations, otherwise the meaning of both may be distorted.

While the printer is typesetting the manuscript in one section of the plant, the art department is handling illustrations in another section so that authors are able to refer to proofs of illustrations as they read and check proofs of text. The original artwork for line illustrations, such as drawings and charts, is photographed, usually at specified reduction, and photoprints are produced. Original glossy prints for continuous-tone figures are photographed through a screen that enables the subtle gradations of gray tones, from white through black, to be reproduced so that, ideally, the resulting halftone photoprints will match the original glossies. Details about reproduction of line art and halftone illustrations are given in later chapters of this book.

Material that will not reproduce well in the printing process may be rejected by either the publisher or the printer, thus delaying publication. When the time constraints or the absence of alternatives require the use of poor artwork, the results will suffer. It is therefore critical that carefully planned and executed illustrations be sent to the printer. Many scientific publications are printed on acid-free paper designed to last 300 years or more. Thus a sloppy illustration once printed may be associated with the author's name for centuries. The printer's objective should be fidelity to the original. When proofs of illustrations are returned to the author and the editor before printing, the author would be wise to review them with the artist or photographer. These proofs indicate the size and quality of the final reproduction, and should be proofread as carefully as the text.

The Author as Consumer

When the printer returns the proof to the author and the editor, he has completed the first phase of his task. At this point the author must assume the role of the reader and look upon the creation in a new way. The text and the illustrations have undergone a transformation, and the author must determine whether the original idea has been communicated properly.

Sometimes the first proofs submitted by the printer are in galley rather than page form. This means that the text and illustrations are not shown in position, but are shown separately. When this is the case, the author and editor make sure that the illustrations do in fact enhance the text, and that they are marked to be inserted at the proper locations. It is then important in the page proof stage to be sure that the merged elements are carefully checked to see that they are in the proper place and sequence.

When proof is provided in page form, all illustrations and text are shown together in their proper relation to each other. The text appears on consecutive pages, with text headings and subsidiary matter such as tables, references, footnotes, and running heads appropriately in place. This makes it much easier for the editor and the author to check the position of the illustrations and to ensure that they are appropriately placed and that they enhance the message of the text.

What about the quality of the illustration proofs? For proper evaluation of the quality of a reproduction, whether line art or halftone, at least one *master* photoprint of each illustration *together with the original artwork* must accompany proof; duplicate proofs of illustrations may be photocopies of the master. But the master photoprints should approximate the final quality of printing, just as the proof of the text approximates the final quality of the printed text. There are four things the author and editor should particularly check in each illustration: (1) legibility of line figures; (2) contrast, particularly in halftone figures; (3) the cropping of the photographs; and (4) the correctness or accuracy of the message.

Legibility

With the printer's proofs in hand, the author should check them against the original. Because an illustration passes through so many stages, it is possible for letters to fall off, scratches to appear, or detail to be otherwise impaired. The more an illustration is handled, the more it is likely to need some retouching. It is the last chance to check the accuracy of the labels, curves, and other data. Equally important is inspection of the master photoprint for quality. Is the focus sharp? Are lines strong and unbroken? If reduction of an illustration has been necessary, has detail been lost? When checking the proofs of graphs, look at every line, arrow, and other symbol to be sure it is correct, intact, and in the right place. Can the words on a graph be read easily? Can the details be seen? Has the change in size altered the consistency within a series of illustrations? Does the illustration fit exactly within the page or column dimensions?

Any problems or corrections should be marked and proof returned to the editor for discussion and correction.

Contrast

The contrast of both continuous-tone and line art should be examined to make sure that the important detail is apparent. In halftone proofs, are the details clear, and does the photoprint compare favorably with the original photograph? Although the screen process precludes an absolute match of the master with the unscreened original, a good printer can produce a photoprint which, to the eye, is a faithful reproduction of the original glossy. If the master photoprint is fuzzier, flatter, or more washed out than the original, the author should circle such areas on the photocopy and specify clearly how the photoprint does not match the glossy. The production staff will convey the appropriate message to the printer and request new proofs. Discussion with the editor and publisher can then identify ways in which the photoprint may be improved.

Cropping

Cropping is a method of blocking out nonessential information at the top, bottom, or sides of a photograph. The editor or publisher may often ask the author about where to crop so that the key parts of the illustration will be retained; photographs can then be shot at a size that will fit the publication while retaining the critical detail. Improper cropping may remove significant detail, so proofs should be carefully examined to see that no essential information has been lost.

Patching up the master photoprints of line art seldom provides acceptable correction of camera-ready quality. Although the author may indicate on the proof what changes are required, it is essential that any corrections be made professionally (by the original artist if possible) on the original artwork. The printer will then make new negatives from the corrected artwork. Corrections for continuous-tone figures require new, original photographs from which the printer can produce new negatives to be used in the final stage of assembly; the old negatives should be destroyed by the printer to prevent any mixup.

Correctness

As a final step, the accuracy of the illustration should be checked one last time. Incorrect data in an illustration are no less misleading than incorrect data in the text. It is as important to "reread" the illustration as it is to reread the text; in fact, it may be more important, particularly when an illustration is not original but is being reproduced from another source. If additional information becomes available after an illustration is created, it may be best to redesign the illustration. Achieving consistency between illustration and text, even at the risk of temporary publication delay, is usually preferable to presenting an illustration that says one thing while the text says another. Any changes in the text must therefore take into consideration the effects they will have on the illustrations.

The Final Task

Once the author has made the final corrections and these have been approved by the editor and publisher, the final task of printing begins. Printing freezes the text and the illustrations at a specific moment in time. Change stops and the publication is presented to the world at large. The effort that has gone into the creation and the transformation of the illustrations now is rewarded if the ideas expressed are understood and used by the readers. Success is measured by the ease with which the reader understands the message of the publication. Clarity of vision as expressed in the illustrations leads to communication of thought, which is the goal of scholarly research.

CHAPTER II
ASPECTS OF THE PUBLISHING PROCESS AFFECTING ILLUSTRATIONS

Understanding the final stage of publishing—printing the illustrations—enables authors, illustrators, and editors to plan and create illustrations at the initial stage so that they will most effectively convey the scientific message. This chapter presents four aspects of production that affect the usefulness and quality of the printed figures: (1) vehicles for scientific illustration (the journal and the book); (2) varieties of printed format; (3) paper; and (4) printing and binding.

The Journal and the Book

Journals are usually published as paperback periodicals, appearing in a pre-established format and on a more or less regular schedule. The author or illustrator preparing an illustration for a journal must take into account the format, the paper, and the printing quality of the journal. The format determines the size of the illustration and of its lettering. The paper affects the quality of the printed illustration and the degree of detail that can be shown. The quality of the printing also can enhance or obscure the detail of the illustration. Each of these factors should provide guidance for preparing illustrations for a journal article. The simplest rule of thumb for the author and illustrator should be: follow what you see in print. Innovations may cause problems, and should therefore first be discussed with the journal editor.

The book, unlike the journal, does not have a pre-established design. A book illustration usually need not be planned to fit a particular format. Unless the book is one of a series with an established format, or the publisher chooses to set specific restrictions on illustrations, the book can be designed to fit the illustration, which allows greater freedom for author and illustrator. However, because of this flexibility illustrations in books may be more expensive to produce than those in journals.

Another significant difference between the journal and the book is that the journal is composed of a series of separate articles, by individual authors, with unrelated styles of illustration. A book, on the other hand, is usually conceived as a total unit, with consistency of writing and illustrations. Artwork produced for a book is therefore consistent throughout, whereas illustrations in journals may vary in appearance from article to article.

When a book is created, the design is often determined by the content. When the format, paper, and printing methods are considered, designers often collaborate with the production staff and printers to produce a pleasing, cost-effective design. The choice of page size, type font, and type size, for example, can be made to suit the subject matter and clarify the text. The type of paper on which the book will be printed can also be chosen to best suit the needs of the manuscript and the illustrations.

The book can be viewed as a custom-tailored product in which the content provided by author and illustrator may dictate the format. The journal, on the other hand, is a standardized product in which illustrative content must usually be molded to a pre-established design. Much of what is discussed in this book is intended primarily as an aid to authors and illustrators, as well as editors, of journal articles; the greater freedom of design in book production imposes less rigid limitations on authors and artists than those imposed on contributors of journal articles.

The Printed Format

The overall format of the book imposes a requirement that the illustrator consider the ultimate size of the illustration at the time of its creation. The format includes not only the selection of type fonts and type sizes but also the number of printed columns, the width and depth of the text page, or "typebed" (the space covered by the type printed on the page), and the trim size of the pages. Illustrations in the journal or book are normally designed to fit within the bounds of the typebed. If the publication has one column of type, the text page and the column width are synonymous. If the publication has two or more columns of type on a page, the illustration can be designed for column width or for text page width. This gives the creator of the illustration a range of options.

In addition to considering the width of the illustration when it is to be printed, the illustrator should also consider the depth (height). Most journals and books have a vertical dimension longer than the horizontal dimension. This means that the illustrations can be designed to be deeper (longer vertically) than they are wide. The illustrator can take advantage of this to design an illustration that may fill one column of a two-column format and achieve a pleasing and economical result. The range of normal ratios of book or journal page format width to height is 1:1.3 to 1.5. An illustration following such a ratio should therefore fit well on the page. An author or illustrator preparing illustrations for journal publication should note the ratio of the height of the page to the width and design the illustrations to fit within this pattern.

Occasionally, illustrations are printed to bleed into the margin or off the page of a journal or book. A "bleed" illustration is one that extends beyond the limits of the text page; it may even bleed across the center margin of the publication to any one of the four trimmed edges of the page. That is, the illustration may extend to the trim edge at the top, outside, bottom, or binding edge. A true bleed illustration has no white space between the illustration and the edges of the page.

Bleed illustrations create production problems and must be handled very carefully. Unless they are essential, from the standpoint of design or information, they should be avoided. Illustrations that extend beyond the standard typebed require special attention at every point in production from the designer, editor, printer, and binder. This means greater expense and greater chance for error. Nonetheless, bleed illustrations, properly used, do have a place in scientific and scholarly publications, and can enhance the readability and appearance of the publication.

The trim size of the publication is an aspect of format that also affects the illustration. Journals and books are normally trimmed in three standard sizes: 6 × 9 inches; 6-7/8 × 10 inches; and 8-1/2 × 11 inches. There are also variations from these sizes, depending on the printing method and technology used. An illustration designed without regard for these sizes is likely to fare poorly when reproduced. In the same way, an illustration designed for reproduction in one of these formats and then shifted to another will face certain problems. For this reason, it is useful for the illustrator to consider the trim size of the publication before creating the illustration.

Paper

Paper not only is an increasingly sizable item in the publication budget but can also significantly affect the quality of the printed page. Many variables in design determine the quantity, size, and quality of paper used. Since these variables can have an important impact on the quality of an illustration, they are briefly discussed below.

Weight

Paper weight is based on the actual weight of a ream (500 sheets) of a standard size for a given type of paper. For example, the sheet size for determining weight of book paper is 25 × 38 inches; for covers, the standard measurement is 20 × 26 inches. If 500 sheets of 25 × 38 inch book paper weighs 70 pounds, it is called "70-pound" stock. Fifty to 70 pounds is an average weight for text stock, and 65 to 80 pounds for covers. When papers having weights above or below these averages are used, extra costs are often involved. The weight of the paper affects the amount of the coating it can carry (see below). Sixty-pound papers and below generally have only one layer of coating per side. Seventy-pound papers and above may have two or even three layers of coating per side. The amount and type of coating affect the quality of illustration that can be printed on the paper. Thus, heavier papers (70-pound and above) are often preferred for fine-quality printing, even though they are more expensive than lighter papers (60-pound and below).

Coating

The distinction between coated and uncoated paper is very important to an understanding of the printing of illustrations. When viewed under a microscope, *uncoated paper* reveals a rough surface of exposed fibers (Fig. II–IA). When ink is applied to uncoated paper three things happen: (1) the ink is applied unevenly because of the uneven surface of the paper, yielding an imperfect result; (2) light is reflected unevenly from the surface, making the illustration appear less sharp; and (3) the exposed fibers tend to absorb the ink and cause it to spread, resulting in a loss of detail and contrast. The pigments in the ink tend to scatter, causing black to appear gray and colors to become dull.

When a coating, usually clay or latex, is applied by the manufacturer to the surface of the paper, it covers the exposed fibers and produces a refined and smooth surface. This has three beneficial results for illustrations: (1) the coated surface enables ink to be applied evenly, thus giving a more precise image; (2) the smooth surface reflects light evenly, making the detail more apparent; and (3) the coating keeps the ink on the surface of the paper and prevents it from being absorbed by the fibers. Printing on coated paper produces sharper, crisper detail, greater contrast, and more vivid colors than printing on uncoated stock.

Several types of coatings are used on journal and book papers. The most common coated sheets are *gloss-coated* (Fig. II–IB). These have a smooth, highly reflective surface which provides the best printing surface for halftone illustrations. This glossy surface is produced by subjecting the coated paper to a calender machine, which consists of rollers that burnish the coated surface, under heavy pressure, to the degree of gloss desired. Because glossy surfaces appear to enhance the reflection of the printed image, as well as to sharpen the detail and contrast, many publications use gloss paper.

Because some people find that the glare from gloss paper makes reading difficult, paper companies also produce a *dull-coated paper* (Fig. II–IC). True dull-coated paper carries as much coating as gloss-coated paper. It is also calendered, but it does not have as smooth a printing surface as does the gloss-coated paper. The pigments used in the dull-

FIG. II–1: *(A) Uncoated, (B) Gloss-coated, (C) Dull-coated, (D) Matte-coated. Each figure is magnified × 25. (Photographs supplied by Consolidated Paper, Inc., Chicago, IL, and reproduced with permission of the supplier.)*

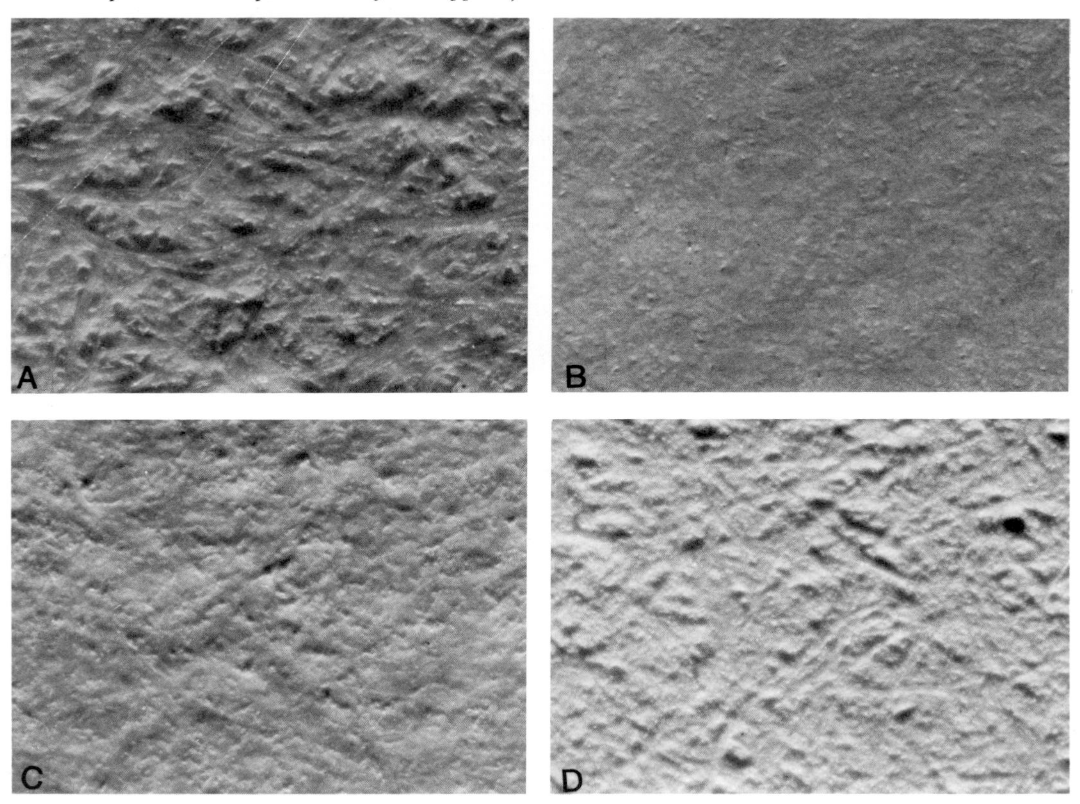

coating process do not burnish to a gloss. Therefore, although the coating keeps the ink on the surface of the paper, the dull surface causes the light to be reflected less intensely and eliminates the glare of gloss-coating. These special pigments, however, are more expensive than gloss pigments. Paper mills therefore tend to charge more for dull-coated paper, sheet for sheet or pound for pound, than for gloss-coated paper.

One alternative to dull-coated paper is *matte-coated paper* (Fig. II–1D). This is a true coated paper which has not been supercalendered. The surface is consequently uneven and dull. Matte-coated paper is less expensive than dull-coated paper.

Another alternative to dull-coated paper is *wash-coated paper*, which does not have as much coating as a true dull-coated paper. The surface is not as smooth and the coating is merely designed to seal the paper fibers to prevent ink absorption. Wash-coated paper allows better printing of illustrations than does an uncoated paper, while costing less than either gloss-coated or dull-coated paper. Wash-coated paper is often used in text books.

Coated papers are more difficult to print than uncoated pages, since the ink takes longer to dry on the coated surface. Coated papers must be run on sheet-fed presses at slower speeds, to prevent the wet ink from transferring or "offsetting" from one sheet to the next. Since web presses run at even faster speeds, special ink-drying units must be added to the press to enable printing on coated stock. If the press does not have such units, only uncoated stock, which allows the ink to penetrate the fibers of the paper, can be used.

Opacity

The degree of print show-through is a variable characteristic of paper. Heavier papers are usually more opaque and have less show-through. In lighter weights, special ingredients can be added in the paper manufacturing process to increase opacity. If paper is not opaque enough, heavy blacks in illustrations may show through from the reverse page and overpower the printed text. An opacity standard of 90, based on a scale of 1 to 100, is regarded as a minimum for a good sheet of text paper.

Brightness

Brightness is a reflective quality of paper. Tones of "white" paper range from a natural ivory to bright white. A black-and-white illustration printed on bright-white paper, rather than on an off-white sheet, appears to have more contrast and depth because of the greater reflection of ink from an intensely white paper. Therefore, for critical halftones such as black-and-white electron micrographs, scanning electron micrographs, and x-rays, or medical subjects with anatomical or microscopic detail, the brightness of paper may be a critical factor in ensuring quality of the printed reproduction. Color illustrations may profit from use of a more neutral-white paper, which tends to balance the colors more accurately.

Permanence

Paper permanence is highly important for some publications. The daily newspaper, the weekly TV guide, and other publications intended to be read and discarded are often printed on the least permanent kind of paper. This paper is often made from groundwood fiber and is highly acidic. Scholarly books and journals, however, often intended for library and archival use, require paper that will last for centuries. This is particularly true where scientific illustrations are used to describe new species in biology. In such cases, the illustration becomes the basic scientific document to which reference will be made over centuries. It is therefore potentially important to the author, illustrator, and editor to know something about paper permanence.

During the period from 700 A.D. to the 1820's, writing, and later printing, were done on pure rag paper. This manufacturing process made paper free from acid and thereby strong and durable. In the late 1800's, with the invention of paper-making machines and steam-driven presses, renewable groundwood fiber gradually replaced rag as the source for paper. As this change took place, acid ingredients, especially alum [$Al_2(SO_4)_3$] were introduced. The greater demand for paper resulted in the manufacture of less permanent papers. Experiments beginning in the mid-nineteeth century suggested that the use of acid contributed to the rapid deterioration of paper. To counteract this problem, many paper manufacturers are now shifting to the manufacture of neutral (pH 7) or acid-free stock for publications intended for extended use. A standard pH of 7.5 is now being recommended, as the residual alkalinity seems to retard deterioration. The shelf-life of poor quality groundwood paper is about one year and of normal-acid paper (pH 5.5 to 6.5) about 40 years, while alkaline-stabilized paper (pH 7.5 and above) may last more than 300 years.

Publishers using permanent, acid-free uncoated paper are now encouraged to use a permanent paper symbol available from the National Information Standards Organization (P.O. Box 1056, Bethesda, MD 20817) in their publications. The symbol is used to indicate compliance with the NISO-developed American National Standard for Permanence of Paper for Printing Library Materials (ANSI Z39.48-1984). A similar standard is now being developed for coated papers.

The problem of paper deterioration among existing books has prompted a number of libraries to seek ways to deacidify already printed paper. The U. S. Library of Congress is currently erecting a large plant at Fort Detrick, MD to handle the massive needs of its holdings. In the private sector, Wei T'o Associates of Matteson, IL offers libraries a non-aqueous spray process for deacidifying printed papers. These efforts emphasize the importance of using acid-free papers to produce scientific works of lasting value.

The Printing Process

In Chapter I, the developmental path of an illustration was described, from the moment of creation through the review and editorial processes, up to the point at which it was approved for the printing process.

Chapter II has dealt with adherence to the journal format and its effect on illustrations, and how the variables in quality of paper can affect the printed results. What remains is to discuss the steps involved in printing and binding the publication, and the effect of these processes on illustrations.

Imposition and Stripping

In letterpress printing, imposition refers to the arrangement, in a particular sequence, of the pages of metal type in a metal press form, called a "chase." A chase contains as many pages as the press can print at one time.

In offset printing, the printer must also plan the sequence of pages that will appear on a single printing plate. Imposition, in this case, refers to the plan for arranging the pages containing text and illustrations. Instead of using metal type, the printer produces photographic negatives of each made-up page. The page negatives are stripped together according to the imposition plan. Stripping consists of affixing to glass or to a sheet of "goldenrod" paper as many page negatives as the press can print at one time. It is at this point that the negatives of any screen halftone illustrations are stripped to the negatives of the pages containing text and line illustrations, where windows for the halftones have been left. Negatives of standing announcements, ads, or fillers may also be inserted to fill any blank pages or blank areas. The imposition process is sometimes done by use of special equipment, such as an Opti-Copy machine, which eliminates the need for manual stripping of individual page negatives into the goldenrod sheets.

Quickly produced prints (blueline, brownline, Dylux, or Van Dyke, for example) are made of each complete signature (a signature consists of a single sheet of publication stock printed on both sides and folded to form a unit of consecutive pages). The folded signatures are then submitted to the publisher for approval; this is the last "proof" before printing. It is especially important at this stage for the production staff and printer to check the identification and orientation of each halftone figure. Unless a serious error occurs, no changes should be made in the blueline proof because of the expense and delays involved in reshooting and affixing the new negatives that such corrections necessitate.

The printing industry will very soon employ machines that eliminate the need for manual insertion of line illustrations into text and the stripping in of halftone negatives. This advanced technology, in reality a part of the typesetting process, includes in the same operation the setting of text, charts, graphs, and line and halftone illustrations. It is made possible by a process known as "scanning" of the artwork and storing of the information in a computer so that it can be combined with the text as it goes to the typesetting machine.

Presswork

After the blueline proof is approved, the negative of each form of composite pages is then "burned" onto a flexible metal plate. The press operator affixes the plate to one of the cylinders on the printing press and then adjusts the flow of ink. With the press running slowly, sample sheets are printed. As the press runs, the printing plate picks up ink onto the image area. The ink is then "offset" or printed in reverse onto the rubber blanket of a second cylinder, which in turn "offsets," or presses, the inked image of the form onto the publication paper. "Offsetting" the image twice in this way creates a right-reading printed sheet (hence the term "offset" press, as opposed to "letterpress," which refers to a printing press on which the paper receives the image directly from the inked metal type and copper or zinc engravings that have been locked into the flat printing bed). The printer examines the sample sheets as they come off press until the ink is the correct color—a term applied to black as well as to other primary colors—and prints evenly across the sheet. When the sample sheets show that the inking is correct, the press is speeded up and the pressrun begins. Periodically during the run, the printer inspects a printed sheet as it comes off press for flaws or inking that is pale or uneven. If necessary, the operator slows or stops the press to clean the plate or adjust the ink flow. Such quality control on press is especially important throughout the run to ensure that printed halftone figures match the photoprints that have been approved.

For color illustrations, a printing plate for each color separation is attached exactly in position on the press. Only exact positioning can assure that each color will print precisely ("register") over the other colors to produce the composite color illustration. If the registration is not exact, the printed figure will be blurred.

Especially pertinent to journal printing at this stage is the order for offprints, commonly called "reprints" (a misnomer). Offprints are printed at the same time that the copies of the publication are printed, whereas reprints are printed by a separate pressrun. Offprints are an inexpensive method for providing an author with copies of an article. The offprint process is also necessary when the article contains color work or fine-quality halftones that cannot be reprinted economically. Offprints of an article often contain extraneous material, especially on the reverse of the opening and closing pages. To avoid such a situation some printers "reimpose" the negatives so that the pages of a single article will not include pages of another article. In producing these "reprints" the printer makes new plates, goes back to press, brings up the ink, and gets out the proper paper—an expensive procedure. Authors are often willing to pay the higher cost of reprints because of the better quality possible.

Binding

A few books and journals have their pages sewn, the strongest and most expensive method of binding. The majority of journals and other paper-covered publications are adhesive bound, an operation that consists of trimming and ruffing the spine edge of the signature, exposing the ruffed edges to glue, and wrapping the cover around the signatures. The adhesive method is often called "perfect" or "tablet" binding. Some journals are side-stapled by a heavy machine that inserts several staples along the spine margins; the spine is then treated with glue and the cover affixed.

Small publications consisting of only a few signatures are often saddle wired, with the pages plus cover opened out at the middle and the staples inserted from the cover spine through to the center of the page spread. For publications requiring frequent updating, a loose-leaf type of binding is efficient and cost effective.

Hard cover books are case-bound. Book signatures are individually sewed, collated, and the spine treated and built up to receive the glue for the hard covers. Soft-bound publications are trimmed at the top, bottom, and outside edges after the covers are affixed; book signatures are trimmed before the covers are put on.

Whatever binding method is used, a small amount of the outer margin is lost in the final trimming. Consequently, when illustrations bleed through the margin, or when text or display matter exceeds the normal page size, it is necessary to check the page proof or blueline proof to be sure no significant matter will be lost in trimming. Similarly, material at the gutter edge of a page must be checked for possible loss, especially when the publication is adhesive bound and the spine is trimmed slightly and ruffed to accept the glue. Sewing or saddle-stitching requires no trimming at the spine, and the pages open out much flatter than when adhesive or side-stapled methods are used. A two-page spread is much more effective when the publication is sewn or saddle-wired.

Case-bound books and large soft-cover publications are often shrink wrapped—enclosed in tight-fitting, sealed acetate—to protect them from rough handling.

After binding, the publications are transmitted to inventory or to the mailing department or a mailing house for wrapping, labeling, and distribution to readers.

LITERATURE CITED

Arnold EC: Ink on Paper: A Handbook of the Graphic Arts. New York, NY, Harper & Row, Publishers, 1972

Blair R: The Lithographer's Manual. Pittsburgh, PA, Graphic Arts Technical Foundation, 8 ed., 1988

Chappell W: A Short History of the Printed Word. Boston, MA, The Stackpole Company, 1971

Craig J: Designing with Type. New York, NY, Watson-Guptill Publications, 1971

Craig J: Phototypesetting. A Design Manual. New York, NY, Watson-Guptill Publications, 1971

Craig J: Production for the Graphic Designer. New York, NY, Watson-Guptill Publications, 1979

Glaister GA: Glaister's Glossary of The Book. Berkeley and Los Angeles, CA, The University of California Press, 2 ed., 1979

International Paper Company: Pocket Pal: A Graphic Arts Production Handbook. New York, NY, International Paper Company, 13 ed., 1983

Lee M: Bookmaking: The Illustrated Guide to Design and Production. New York, NY, R. R. Bowker, 2 ed., 1979

Mundy PJ, Radford HT: Illustration preparation and reproduction techniques. The publishers' needs. Med Biol Illus 26:111–114, 1976

CHAPTER III
PREPARATION OF ARTWORK

Overview

In publishing language, the word "illustration" is used interchangeably to refer to either a drawing or a photograph. The words "graphics," "graphic art," and "fine art" are occasionally used regarding illustrations, but these refer principally to various technical differences in treatment of a subject. The word "graphics" broadly encompasses graphs, charts, diagrams, and the design and use of written or printed characters. The term "graphic arts" includes engraving, etching, lithography, photography, serigraphy, and woodcut, while "fine art" refers essentially to painting and sculpture.

This chapter is concerned with pictorial drawings that are prepared to accompany written information in a scientific publication.

Successful scientific drawings enhance the written word but do not take its place nor reiterate its message, and should not require lengthy legends to be understood. Scientific drawings should clarify concepts, provide visual orientation, and simplify complex information. They must be pleasing to look at, cleanly rendered and accurate.

The author is well advised to seek the services of a professional illustrator and to begin plans for the necessary artwork well in advance of the publishing deadline. When a highly realistic drawing is required the artist can most accurately develop it if the subject is viewed directly. In depicting a surgical procedure, for example, the artist may view the operation and make initial sketches in the operating room. If medical or scientific specimens, instruments, hardware, or materials are to be shown realistically, these should be made available to the artist while the artwork is in progress.

Beginning work well ahead of the publishing deadline will ensure the best results. If a drawing is hastily or amateurishly prepared it will usually prove to be costly, either in a notable loss of credibility for the manuscript, author, and publication, or in the extra dollars required for reworking.

Selection of the medium for rendering may be limited by the publisher's and author's budgets and specifications. In general, black-and-white artwork is preferred, either in line or tone.

Line work is the most economical and predictable art technique to print. It is an excellent medium for charts, graphs, diagrams and simplified drawings.

Tone (also called continuous tone) drawing incorporates the full range of values from black to white and is an effective medium for depicting fine detail.

In some instances one or two flat colors can be overprinted on black-and-white artwork to create visual emphasis. Full color artwork is expensive to print but there are occasions when it is the only appropriate medium for a specific project.

Quality Evaluation

Composition

Good composition determines what is pleasing to look at in a drawing, and in a scientific illustration it also enhances the instructive character of the subject matter. Composition is the arrangement of artwork on the printed page, the balance between the visual components within the artwork, and the relationship of the drawing to leaders, labels, and negative (empty) spaces. The artist is responsible for arranging information within the illustration and rendering it in a way that most clearly conveys the message. Composition must be carefully planned in order to maintain logic and understanding as well as visual appeal.

Ideally, there should be a balance between form and space, between areas that are filled with detail and areas containing less for the eye to absorb. The reader's eye should be led readily to the central message of

FIG. III–1: *A. Example of poor composition. Central message is lost due to apparent lack of planning for placement of all elements. B. Central message is more clear. Labels and leaders are balanced with artwork. (Reproduced, with permission, from Way, LW (ed.): Current surgical diagnosis and treatment. Los Altos, CA, Lange Medical Publications, 6 ed., 1983.)*

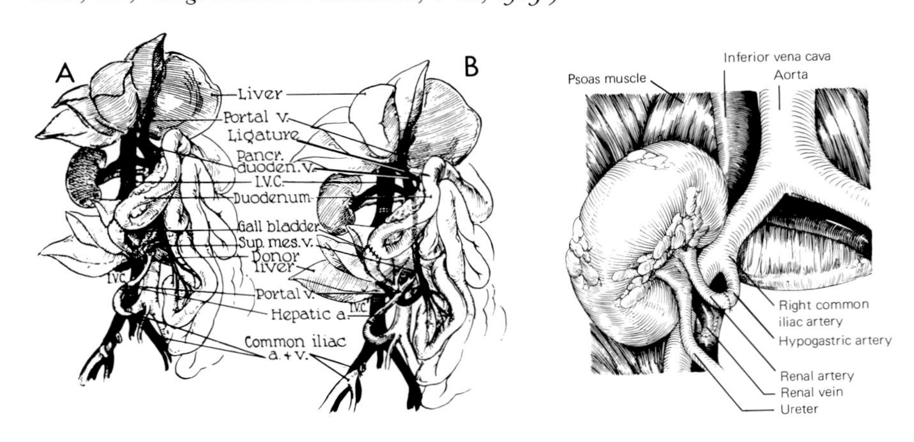

the work. The main area of the drawing should contain the most detailed rendering and forms that are fully depicted. Other elements, such as insets and labels may be included for orientation and completeness of information, but the central focus of the illustration should not be obscured by this secondary material.

Light Source

The assumption of an imaginary light source shining from the upper left corner of an illustration is a traditional convention that enables the artist while creating a drawing or painting, to predict where light and shadow should fall. This light source is a constant, and is applied to all elements of an illustration. Understanding its effects will dictate the artist's development of contrast, reflection, perception of height or depth, opacity or transparency, and modeling.

Clean Copy

Line artwork prepared for high contrast reproduction must have the greatest degree of contrast possible between the black lines, the dark areas and the ground color. The background should be white, or as near to white as possible. A transparent ground should be mounted on a white backing board. Any pencil sketch lines, smudges, or fingerprints must be removed before reproduction. Lines and solid areas in the drawing must be dense black, with all edges and corners sharp and clean. Lines that are fuzzy or gray, and edges of lines or areas that are not crisp, will be unpredictable and usually imperfect in the final reproduction.

Tone drawings must be free of fingerprints, sketch lines, tears, and tape marks, since these will be recorded by the printer's camera as faithfully as the drawing itself.

Overlays for spot color must be accurately in register with the base drawing. If there is a discrepancy in the match of the register marks, the color spot will not fit within the artwork when it appears in print.

Full color illustrations require the same care in preparation and handling as line or tone art. They must be clean and free of artifacts.

Specifications and Proofs

At the beginning of a project the artist should be provided with information about page size, image area, and column width in the proposed publication so that the art can be prepared in the correct proportion for these specifications. The artist should also be informed of the space allocated by the publication for artwork, as this will serve as a guide in designing the art to fit one column, a full page, a half page, and so on.

If the artist is to affix labels to the illustration this should be specified, as well as the typeface preferred by the publisher. If there is no preference then the artist should be informed of the typeface used in the printed text so that a complementary alphabet can be selected for the labels.

Whenever possible, the artist as well as the author should be provided with proofs so that any necessary modifications in reduction, contrast, positioning, and so forth can be requested.

Margins

There must be adequate space or margins around the image areas and beyond the labels of any illustration. These permit the printer's camera some latitude on the printing negative, allow room for register marks to be placed away from the rendering, and prevent handlers from touching the actual surface of the drawing.

Reduction and Sizing

Drawings in any medium should be prepared 50% to 100% larger than

FIG. III–2: *A. Proportional scale (also called a scaling wheel), is used to obtain accurate dimensional ratios. B. Artwork appears smaller as reducing lens is moved away from it, thus providing a quick check for clarity of detail after reduction.*

A

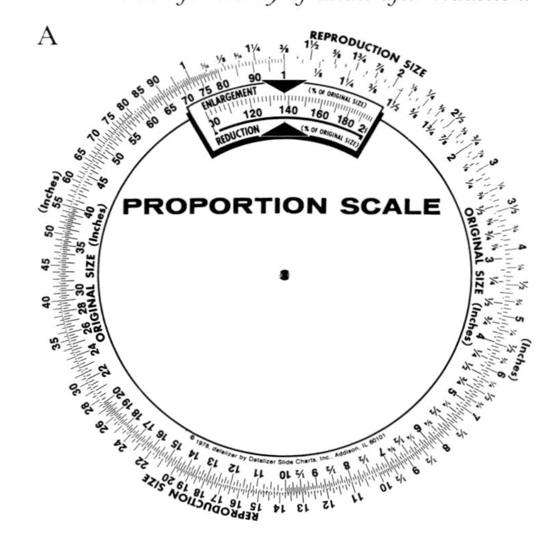

B

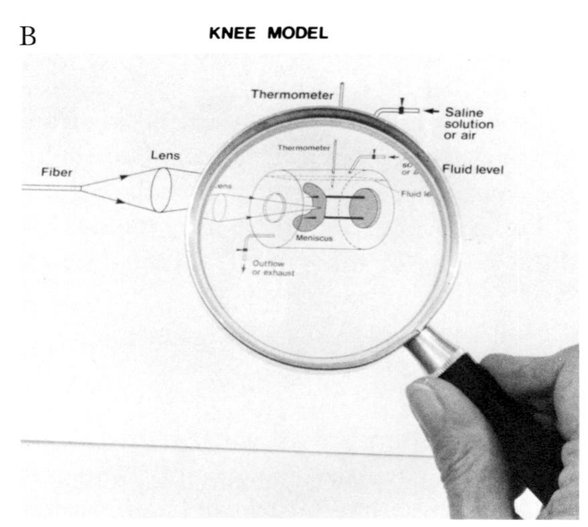

FIG. III–3: *The effects of reduction. A. Original size (100%). B. 67% of original size. C. 35% of original size. (Polycystic kidney. Artist: Ralph Sweet. Reproduced, with permission, from: Smith, DR. General urology. Los Altos, CA, Lange Medical Publications, 11 ed., 1984.)*

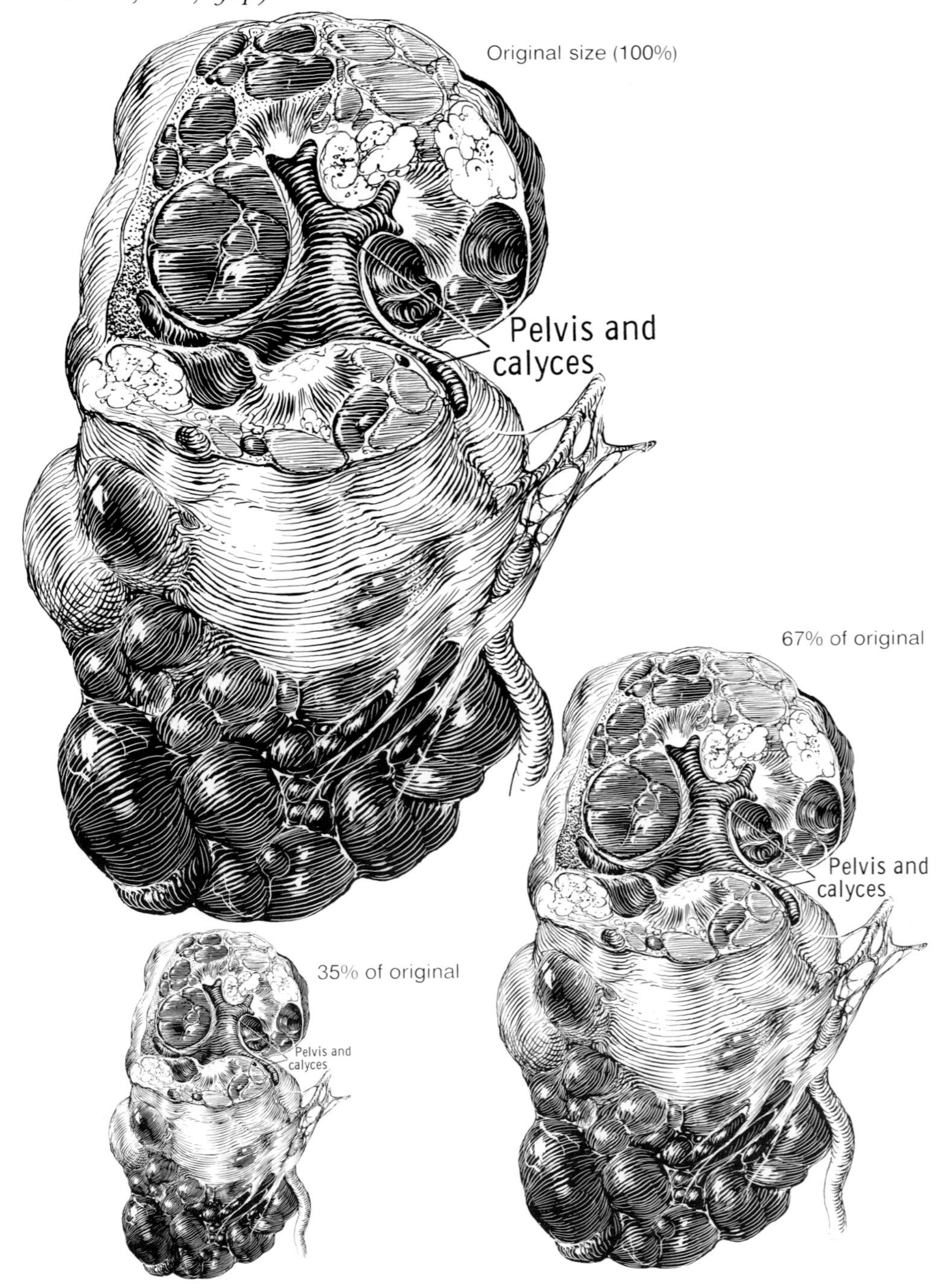

their final size on the printed page. They are then reduced in print to 67% or as much as 50% of the original size. Knowing the publication's format enables the artist to plan the illustration, including the area required for labels, so that it will reduce proportionally to fit the prescribed space. The artist will need to know the text page size without running head and folio, number of columns and their widths, and how many illustrations are required in the manuscript. The illustrations are then designed to fit within one column, half a page or a full page after reduction. A proportional scale simplifies figuring the correct dimensions of the reduction ratio for artwork. A reducing lens is also a useful tool that quickly shows how the illustration will look in various degrees of reduction.

Today many xerographic copy machines can reproduce an image with a given ratio of reduction or enlargement, which can be selected by the operator. If this type of equipment is available it offers a useful visual guide for fit, but tends to distort contrast and sharpness. Xerographic copies should never be submitted for reproduction in print.

Instructions for reduction are written for the printer in the lower margin of the artwork, out of camera range. The instructions should

FIG. III–4: *A simple method of determining the reproduced dimensions of a drawing in relation to its original size. This type of scale can be sketched before drawing is begun, to establish working size.*

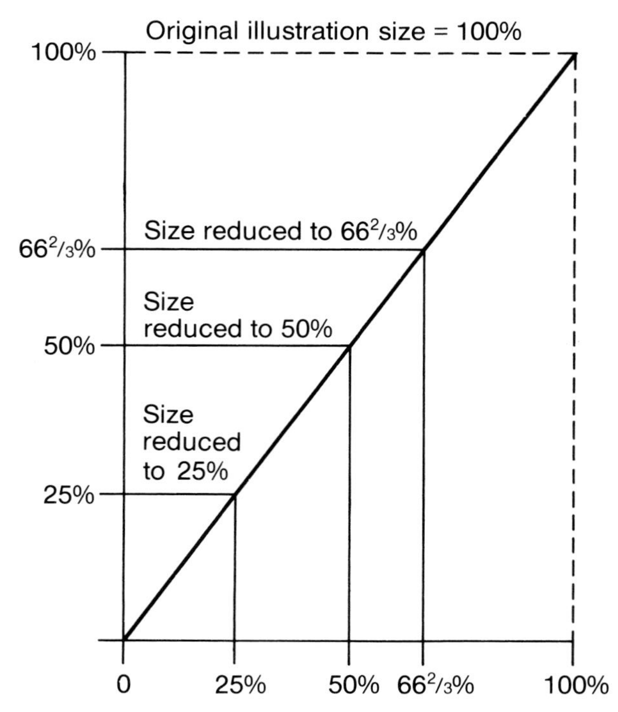

FIG. III–5: *Comparative effects of reduction and enlargement.*
A. Original size. B. Reduced to 67% of original size.
C. Enlarged to 300% of original size.

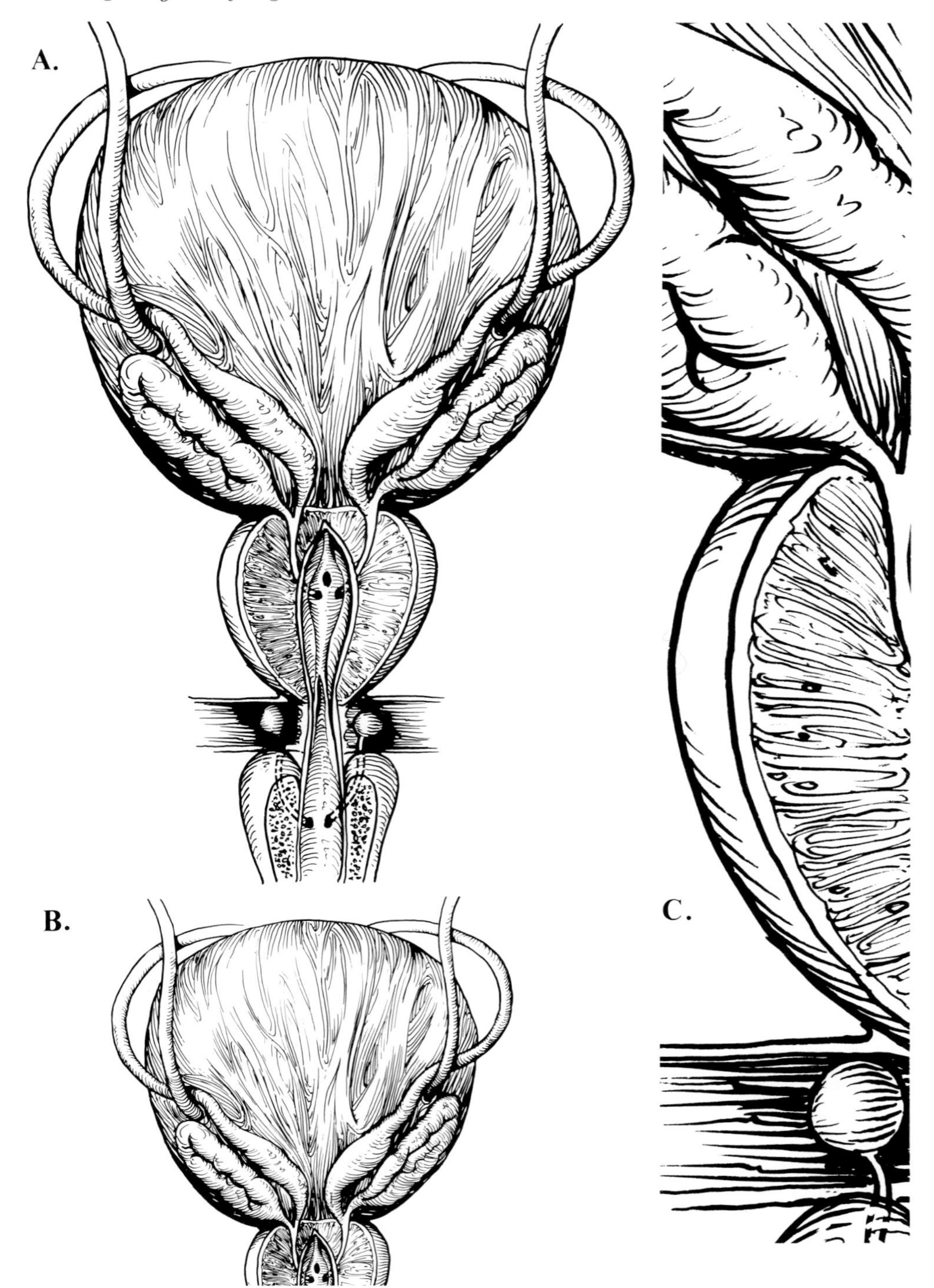

read, "Reduce to 67% (or 50%, or whatever percent is required) of original size."

Reduction tends to smooth out inconsistencies in a drawing and to some degree may serve to sharpen line edges. However, reduction can never save a poorly executed drawing. Reducing an illustration much more than 50% of the original size usually causes light areas to weaken or possibly to disappear entirely, and dark values to block up or lose form. Over-reduction of all types of illustrations should be avoided.

A line drawing should never be enlarged for reproduction. Even though the work may look very sharp to the naked eye, enlargement emphasizes small flaws, rough surfaces, and fuzzy edges.

Line Art

Line art is black and white, with no shades of gray. It consists of black lines or portions of lines, such as dots and dashes on a white background. It may also be reversed to appear as white on black.

Ideally, there are no grays in line copy. If gray areas are included in a line drawing their reproduction may not be predictable. The printer may try various photographic techniques to increase contrast in an at-

FIG. III–6: *A. Positive: black on white. B. Negative: white on black. (Artist: Pieter Folkens. Reproduced with permission of the artist.)*

A

B

FIG. III–7: *A. Line drawing with grays added for shaded effect.*
B. Grays removed.

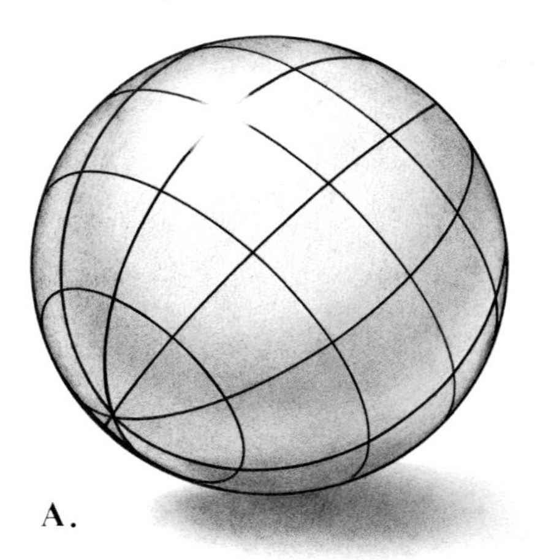

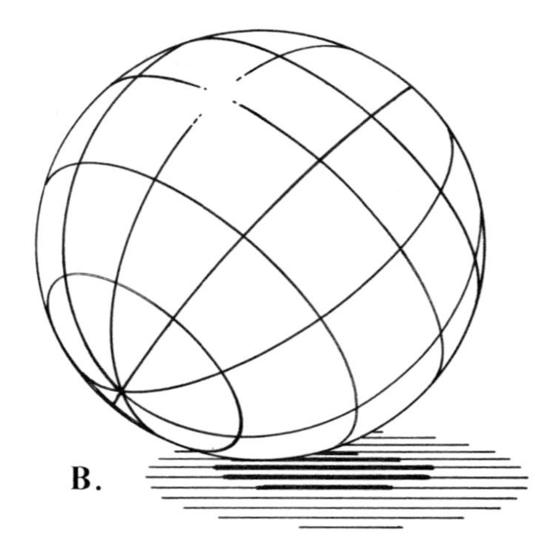

tempt to hold the grays, but this can also intensify the true blacks and give an undesirable effect. In general, any illustration in which black-and-white artwork is combined with grays will not look as sharp in print as in the original, unless special production techniques are used (see page 69, Double Image Prints).

Types of Line Art

A good pen and ink drawing is pleasing, informative, economical to print and reproduces well, even on poor grades of paper. It is probably one of the most common types of scientific artwork in use today.

Most line illustrations are rendered with black India ink, pens, or brushes. Technical pens give an evenly weighted line and are sold in 12 different widths. Dip pens have either stiff or flexible nibs. Flexible nibs produce lines of varying widths, with thick and thin gradations that depend on the pressure applied by the artist. A brush with ink is often used to create smooth, tapered lines, whereas a drybrush is used to produce fuzzy effects.

Line Tone and Texture

An illusion of tone can be created by combining lines or portions of lines. Placed close together, these produce the effect of dark tones, and placed farther apart they give a lighter value.

FIG. III–8: *Tools used for pen and ink rendering. A. Technical pen. B. Flexible pen. C. Brush. D. Dry brush.*

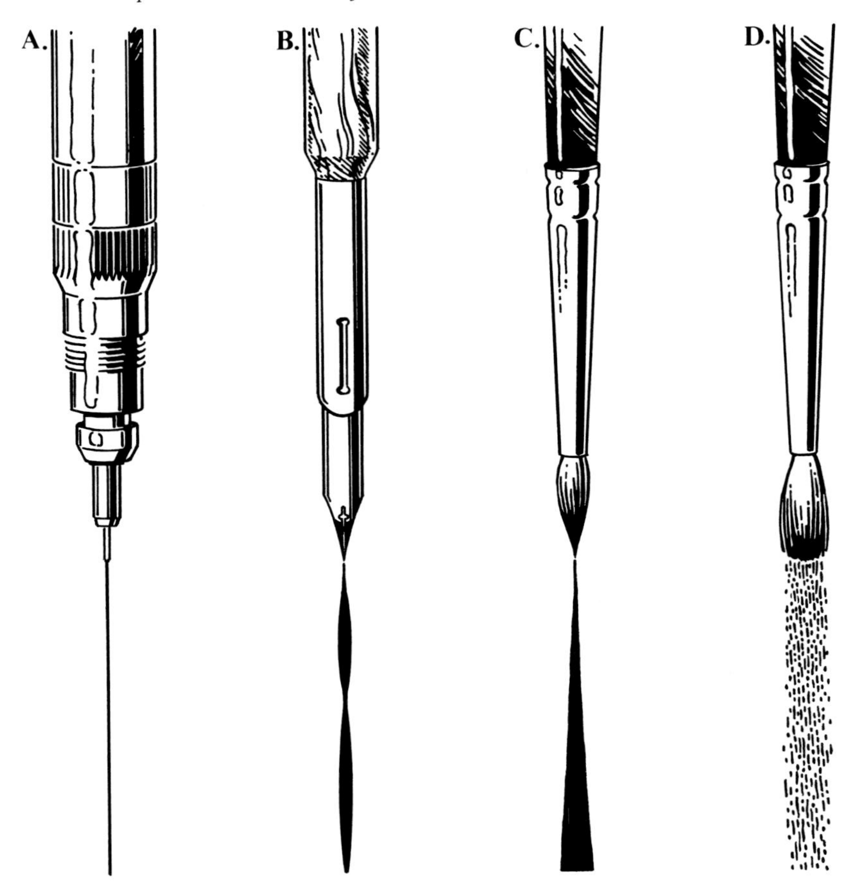

Stippling

The smallest part of a line is a dot, also called a stipple. Stippling is a common technique used to create values and form in line art. It is easy to control and time consuming to produce but generally gives an accurate depiction of subtle variations in form, light and shadow. Stippling is applied with either a rigid or a flexible nib, although the technical pen produces a more consistent size and shape to the dot.

Cross-hatching

Cross-hatching is achieved with intersecting lines, and also establishes tone and form.

Contour lines

A good line illustration by an experienced artist will appear to flow without stiffness and will effectively depict texture, contour, depth, and

FIG. III–9: *Stippling used to create form, detail, and values. (Reproduced with permission of the artist, Keiko Hirstsuka. Collette BB, Russo JL: A new species of Spanish mackerel from Australia and New Guinea. Aust. J. Marine Freshwater Res., 31:241–50, 1980.)*

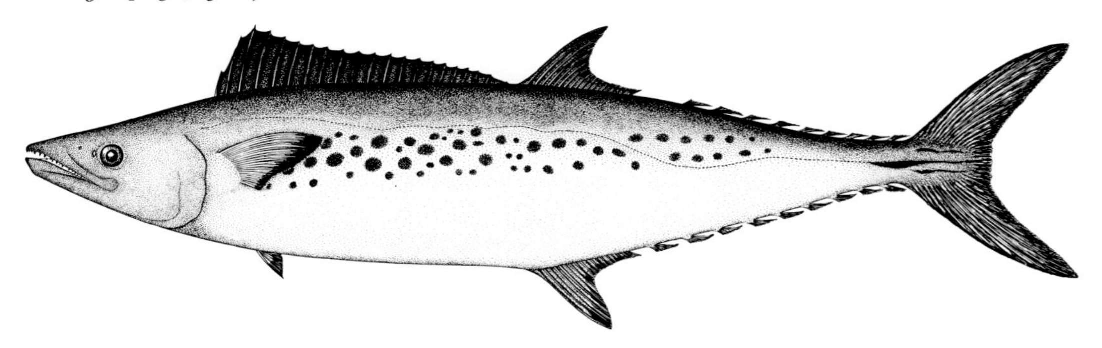

detail. A complex pen-and-ink illustration often contains a variety of lines, including contour lines, sometimes called eyelashing. These are drawn with a flexible pen and are thick and thin lines that either parallel or lie perpendicular to the edges of a shape; by their variations in weight they depict tone and form. This is a difficult technique, and if not well executed will be totally ineffective. Heavy lines with the proper space between them visually bring an area forward, and as the lines become thinner and closer together the edges of a form appear to fall away. On occasion, a drawing may be too fully rendered, with insufficient white areas for highlights and contrast, and the impact of the message is lost. However, well-drawn contour lines can give the effect of an engraving.

Both cross-hatching and contouring must be carefully done on drawings of biological specimens (plants or animals), since the lines occasionally can give the impression of nonexistent texture, pattern, or structure. Frequently, several styles of rendering are combined in a line drawing.

Loose pen and ink rendering

It is not necessary for a carefully rendered drawing in pen and ink (or any medium) to appear static or over controlled. The creative solution for depicting movement, for example, may employ repetitive forms and lines that emphasize directional flow and changes in position, rather than the precise details that are more visible on a stationary object. A familiar example of this approach is the 1912 cubist painting of "Nude Descending a Staircase" by Marcel Duchamp, and we all recognize the cartoonist's techniques for depicting motion. The apparent sketchiness of this type of drawing does not in the least imply carelessness on the part of the artist. In fact, only the artist's close attention to placement of the object, the subtle changes in balance and direction, and restraint

FIG. III–10: *Crosshatching used to depict form, light, and shadow. Note the stippling used to create the image of calyces beneath kidney surface. Abnormal right kidney with double ureters and pelves, and accessory renal vessels. (Artist: Gerald P. Hodge. Reproduced with permission of the artist.)*

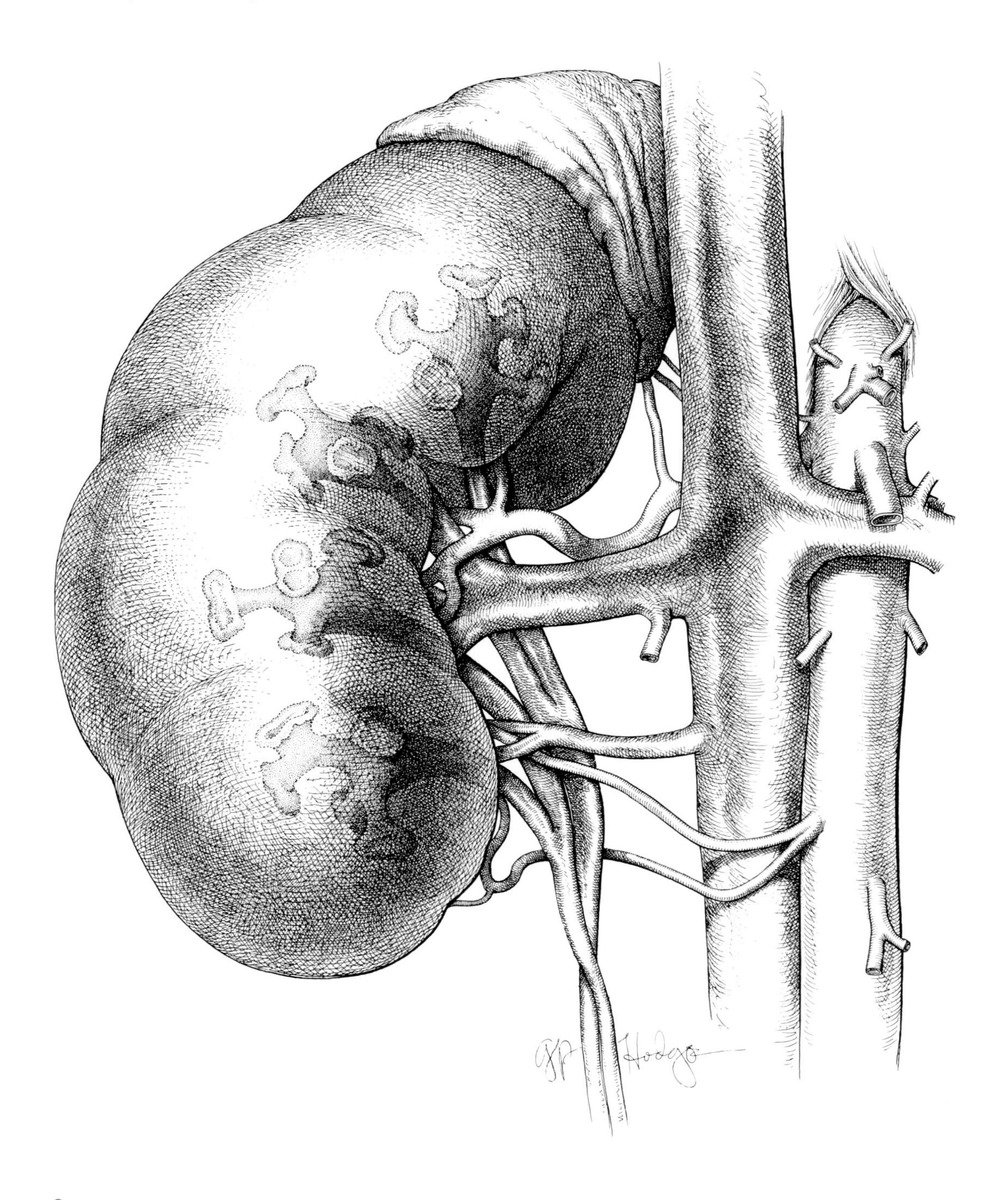

FIG. III–II: *Contour lines (eyelashing) used to create form, light, and shadow. Note that changes in direction of the contour lines make the object appear to move in or out of the drawing. Ileosigmoidostomy, by Russell Drake. (Reproduced with permission of the Mayo Clinic, Rochester, MN.)*

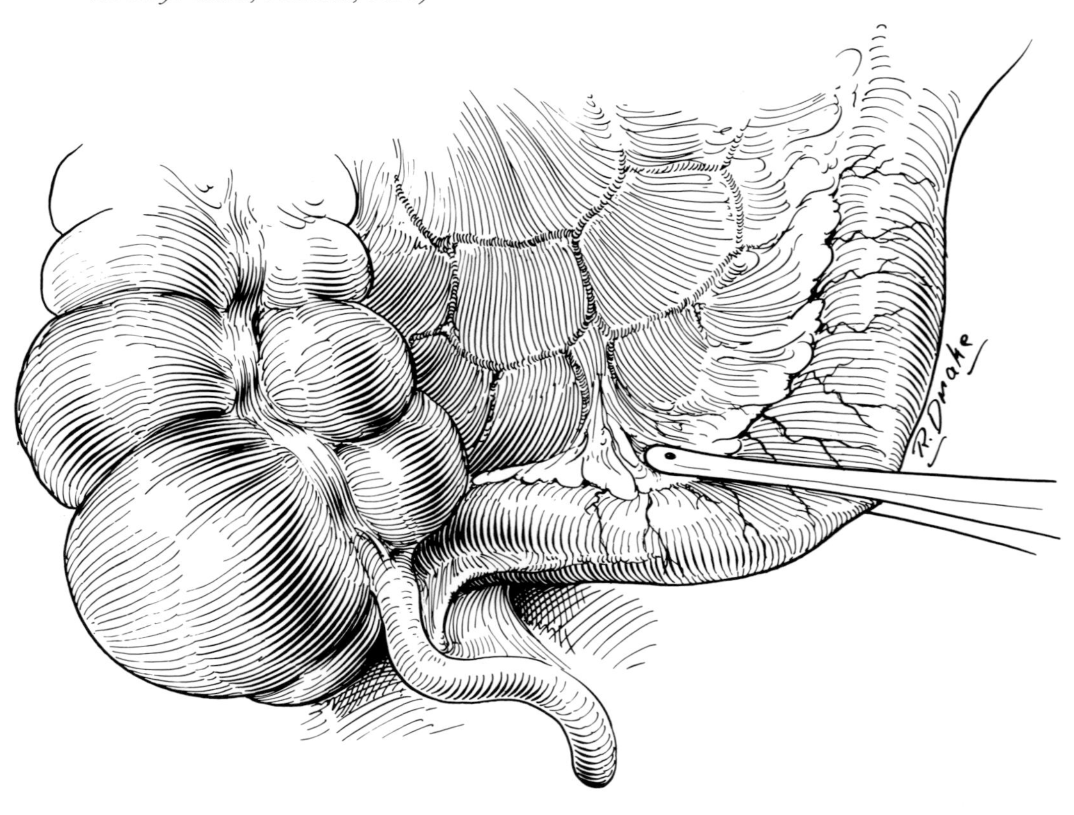

in rendering can create the effect of movement and produce a free and loose illustration.

Coquille board

Texture can also be produced on a roughly textured surface, such as Coquille board, by rubbing it with charcoal, conte, black pencils, lithographic grease pencils, or even felt-tipped pens. The result gives the effect of ink stippling yet it can be produced very quickly. Ink is generally used for the outlines and linear aspects of such a drawing, although definition of fine detail on a textured surface is difficult. If the medium is used too lightly the resulting gray areas in the drawing will drop out on reproduction, which can jeopardize the visual message. For this reason, Coquille board drawings often reproduce better as halftones.

FIG. III–12: *Contour lines, crosshatching, and stippling combined, illustrating the configuration and values of the face as well as the procedure beneath the surface. (Artist: Carol Donner. Reproduced with permission of the artist. From McCredie, JA (ed.): Basic surgery. New York, MacMillan & Co., 1977.)*

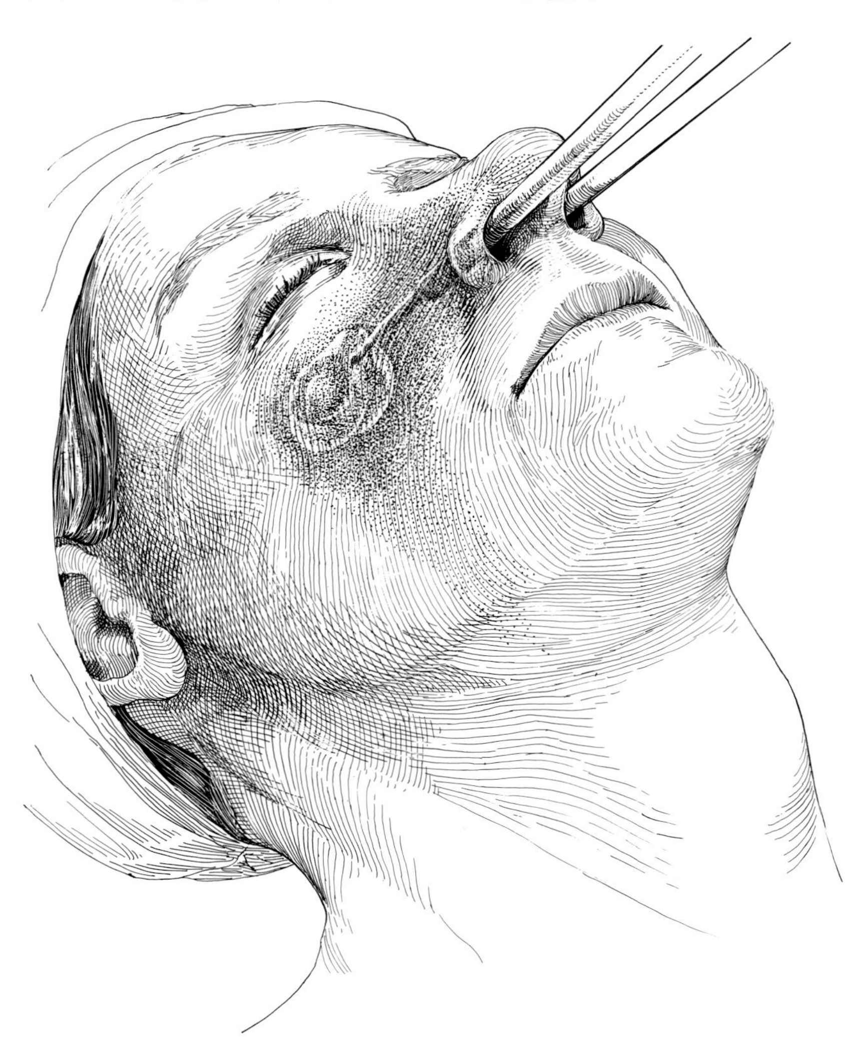

Preprinted Line Screens and Patterns

Instant tones can be created on line copy with screens or patterns. These are commercially manufactured materials which are also called shading mediums, Benday patterns, or line tints. Such preprinted patterns are great time-savers in preparation of maps, graphs, or drawings that re-

FIG. III–13: *Line is used freely and loosely to convey movement. (Artist: Gottfried Goldenberg. Reproduced with permission from The Archive of Medical Visual Resources, The Francis A. Countway Library of Medicine, Boston, MA.)*

quire large areas of uniform tone or texture. They are slightly adhesive, printed films mounted on a translucent backing and come in many designs, patterns, and textures.

A piece of the selected film is peeled from its backing and placed on top of the drawing. The portion needed is then cut out with a sharp blade following the outline of the drawing below that is to be textured. The leftover film is lifted off and the film on the drawing is burnished down for maximum adhesion. If small air bubbles appear after bur-

FIG. III-14: *Creation of the effect of stippling on Coquille board with ink and pencil.* Actinogyra miihlenbergii. *(Artist: Lucy C. Taylor. Reproduced with permission of the Smithsonian Institution, Washington, DC.)*

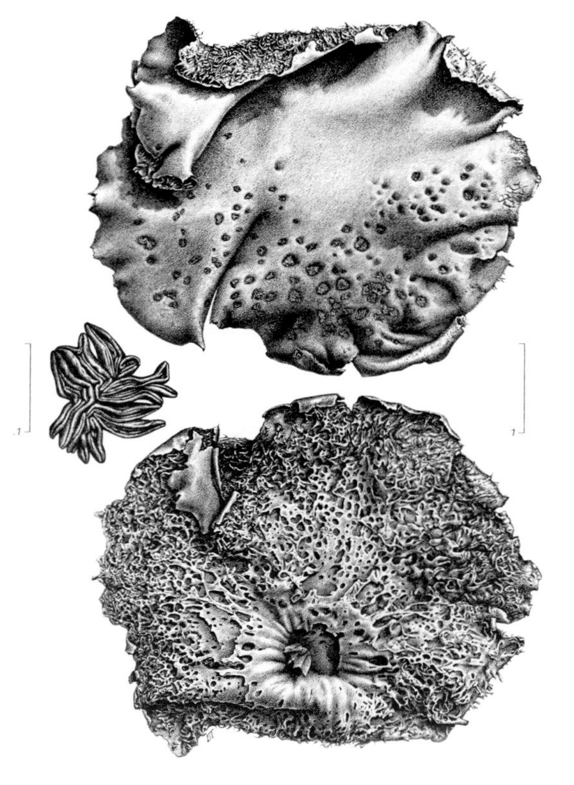

nishing, these can be carefully pricked to release the air and the film then is pressed flat. It is important to cut the film so that it fits exactly the area covered and is not left ragged along the edges. When a film is cut against a black outline, the knife may occasionally lift out small bites of the ink. These white spots should be carefully touched up to ensure a clean black line.

Some screens and patterns are rubbed onto the drawing much as is done with transfer type, but this material may chip off if the original is not handled with care.

Adding Color to Line Art

Line art that has been prepared for publication is frequently expected to serve double duty as a color slide or exhibit piece. In such instances

FIG. III–15: *Preprinted patterns applied to artwork to differentiate areas of sensory distribution. (Artist: Laurel V. Schaubert. Reproduced, with permission, from Way, LW (ed.): Current Surgical Diagnosis and Treatment. Los Altos, CA, Lange Medical Publications, 6 ed., 1983.)*

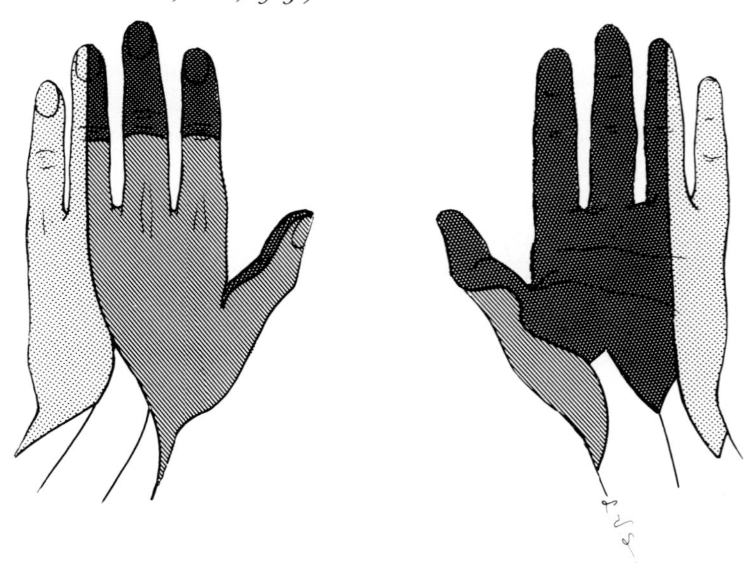

the line art can be drawn on translucent or transparent material and placed on top of a colored board, or it may be back-painted.

The Color-Key process is another option for converting all components of a line drawing to opaque or transparent color. This process involves the use of ultraviolet light and chemicals that can reproduce the drawing in a variety of colors on a clear acetate sheet (see Chapter VIII, page 226).

Tracings

Tracings made for diagnostic study of body functions such as electrocardiograms, electroencephalograms, etc., are occasionally submitted as illustrations. The paper used for these tracings is often pressure sensitive, and therefore over-handling it will produce additional permanent black marks and artifacts. Whenever possible, tracings should be photographed on high-contrast film and the print trimmed and mounted, thus avoiding damage to the originals.

Lettering

Lettering included in any scientific drawing, whether it is a diagram, a chart, a graph, or a complex illustration, must be considered an integral part of the design and composition of the entire piece. The placement of the lettering, its balance, and its relative importance are planned at the time of the initial creation to fit and complement all other elements of the drawing.

Some drawings, such as diagrams or flow charts, are created to display a sequence of events or a hierarchy of ranking and may be composed mainly of words and directional lines. Graphs can contain lettering that defines the measurements and statistics, and illustrations usually include labels that describe or identify various important structures in the picture.

A basic understanding of the fundamentals of good letter design, spacing, and optimal application techniques is important for anyone who is planning, creating, or evaluating a scientific illustration.

FIG. III–16: *Examples of type sizes from 6 to 60 points.*

6 abcdefghijklmnopqrstuvwxyz
7 abcdefghijklmnopqrstuvwxyz
8 abcdefghijklmnopqrstuvwxyz
9 abcdefghijklmnopqrstuvwxyz
10 abcdefghijklmnopqrstuvwxyz
11 abcdefghijklmnopqrstuvwxyz
12 abcdefghijklmnopqrstuvwxyz
14 abcdefghijklmnopqrstuvwxyz
18 abcdefghijklmnopqrstuvwxyz
24 abcdefghijklmnopqrstuvwxyz
30 abcdefghijklmnopqrstuvwxyz
36 abcdefghijklmnopqrstuvw
42 abcdefghijklmnopqrstu
48 abcdefghijklmnopq
60 abcdefghijklmn

Alphabets

Each letter in an alphabet is called a character. Capital letters are called upper case or caps. Small letters are called lower case; some consist of a body and an upward stroke (ascender) and some of a body and a downward stroke (descender). A font is a complete assortment of a given size and style of type, containing all the characters for setting ordinary composition.

Type size is measured from the top of an ascending letter to the bottom of a descending letter. The body, or x-height, makes up the greatest portion of a letter. The scale for type measurement is in points. There are 72 points in an inch. The length of a line of type is measured in picas. One pica equals 12 points (1/6 or 0.166 inches; 0.42 centimeters) and 6 picas equal 1 inch (2.54 centimeters). Type is usually available in 6 to 72 points.

Alphabet Design

The design of an alphabet is called face or style. Examples are Helvetica, Caslon, and Univers. Type is designed vertically (roman), slanted (italic), with straight strokes (sans serif), or with a cross stroke at top and bottom (serif). Most alphabets are also available in different weights, such as light, medium, or bold.

Readability and Legibility

Most people who are not concerned with the fine points of typography use the words "readability" and "legibility" synonymously. However, there is a distinct difference. Readability is the ease with which type can be read, whereas legibility refers to the speed with which each letter or

FIG. III–17: *Anatomy of letters. Note that curves at the base of rounded letters descend slightly below the baseline.*

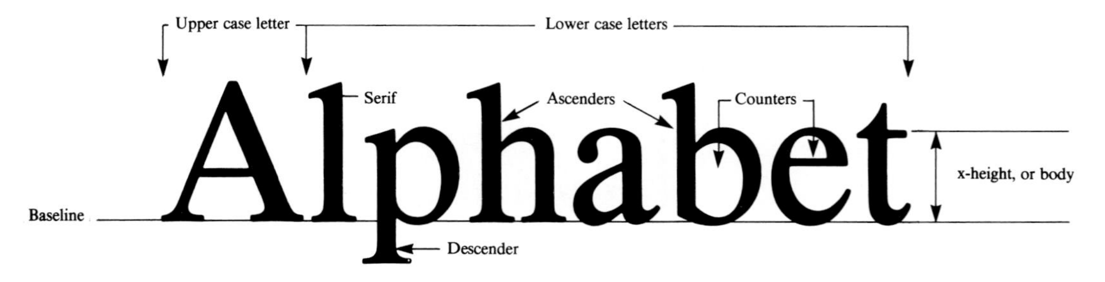

FIG. III–18: *Comparison of inch and pica measurements.*

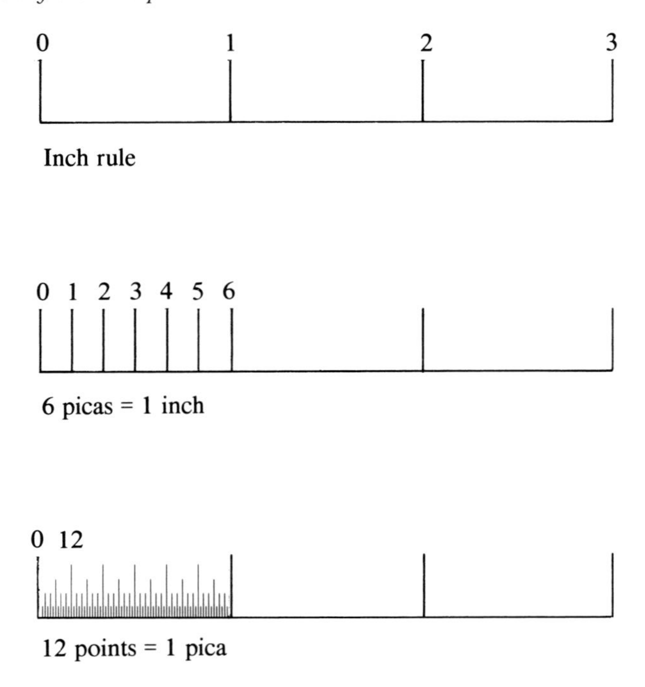

FIG. III–19: *Examples of alphabet designs and weights.*

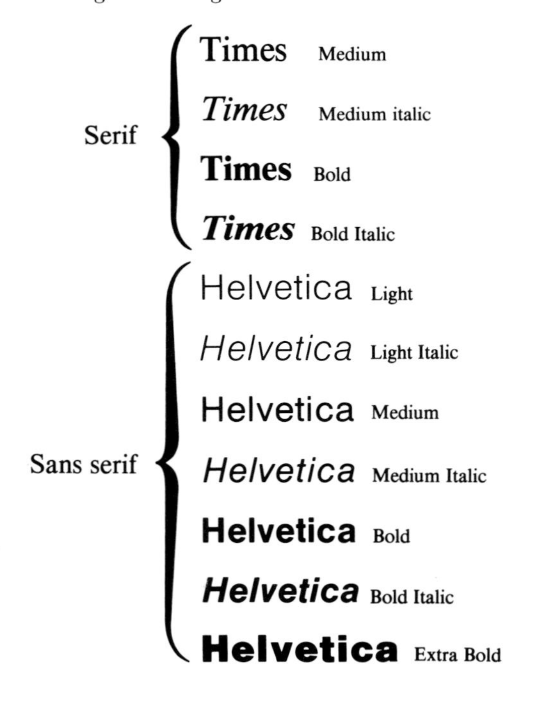

word can be recognized. Readability refers to type arrangement; legibility is concerned with type design.

Readability and legibility depend upon several factors that must be considered when selecting a typeface. These include paper texture and finish, ink color, type size, line length, line spacing, etc. Type should be set so that it can be read with little effort or eyestrain. Proper line spacing is important to the appearance of any printed piece. Each project presents different problems, depending on the type used, as, for example, whether all capital or all lower-case letters are to be used, or whether italics are to be included. An artist with skill in type design will make sure that the spacing enhances the typography. In practice, good spacing is often a matter of good artistic sense.

Guidelines for Success with Type

A few basic pointers for the successful use of type may prove helpful.

1. Type must be legible and evenly aligned.
2. Conventional typefaces are preferred in scientific illustration (for example, Helvetica, Univers, Futura, or Times).
3. Leading (space between lines) must be sufficient to permit easy reading and prevent crowding.
4. Upper/lower case copy is the easiest to read.
5. All capitals should be reserved for titles or headings.
6. Italics, rather than underscoring, should be used to create emphasis.
7. Characters should be limited to no more than 75 per line.
8. Type should not be overprinted on a drawing or on a textured surface.
9. Anyone preparing to use type should learn to use type charts, to ensure accurate specifications for the typesetter or printer.
10. Never copy the latest fad!

Application of Lettering

A wide variety of typefaces and point sizes are available on preprinted transfer sheets. These letters are easily transferred to most surfaces by simply placing them in the desired position and rubbing each one with a smooth object, such as a burnisher, until it is separated from the carrier sheet. If a mistake is made the letter is easily removed by gently scraping or by lifting it with a sticky surface such as tape or a rubber cement pickup.

Transfer type is used for simple character-by-character composition

FIG. III–20: *Application of transfer type.*

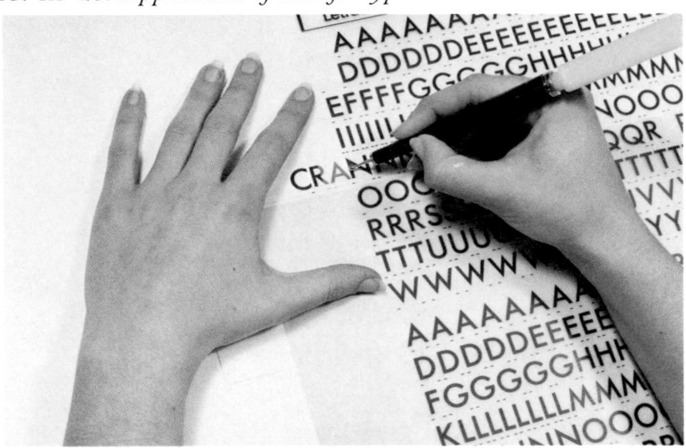

of words and lines for headings, labels, or displays, as is cut-out lettering.

Lettering machines are hand operated to produce type, letter by letter, that is then transferred to the artwork.

The Kroy machine employs type discs which contain various alphabets in different point sizes. These discs are dialed to the correct position and each letter is pressed onto a tape. The tape is either opaque or transparent and is affixed in strips to the artwork.

The Leteron machine generates pressure-sensitive type in various colors that are applied letter by letter to the ground. This system is best suited to display rather than publication.

Several other brands of lettering machines are available which operate similarly to produce letters made of a variety of materials.

Phototypesetting provides type that is photographically set, one character at a time on photographic paper. The paper is cut by the artist and affixed to the artwork or an acetate overlay. This procedure is called "paste-up."

Phototypesetting (photocomposition) is a sophisticated process with practically no limitation on type sizes or alphabets. Phototype is clean, without irregularities, and can be set as tight or as open as desired.

Today most phototypesetting equipment is computerized. Two types of systems are customarily employed: (1) *digital* typesetters, which use fonts stored electronically within their memory chips, and (2) *analog* typesetters, in which the fonts are actual photo images stored on interchangeable discs. A wide variety of fonts and symbols are available with either system.

Using a keyboard, the typesetter operator creates copy which is displayed on a video screen and automatically stored in the computer's short term random access memory (RAM). The copy can also be trans-

FIG. III–21: *Examples of phototype.*

B

PHOTOTYPE

Typesetting
Typesetting

ADJUSTABLE LETTERFIT

ferred to a magnetic diskette for long-term storage, and can be output to photosensitive paper either from RAM or the diskette.

After the photosensitive paper is exposed to the copy within the computer, it feeds into a light-proof cassette. The cassette is then loaded into a processor which unrolls the paper and develops the image, fixing, washing, and drying the paper in minutes. The resulting galley proof is then ready for paste-up.

Freehand lettering by any method requires skill and practice. When attempted by an amateur the work is seldom acceptable.

Lettering can be produced with technical pens which come with rigid tips in various widths. These can be used with either a stencil or a lettering device, such as Leroy, Wrico, or Varigraph. The technical pen is mounted in a scriber. A stylus in the scriber fits into the grooves of cut

FIG. III–22: *A. Example of lettering produced with Leroy scriber and template. B. Leroy scriber and template in use.*

ABCDEFGHIJKLMNOPQRSTUVWXYZ&(%)!;
abcdefghijklmnopqrstuvwxyz 0123456789·

FIG. III–23: *Example of typewritten alphabet produced with "Title" ball on an IBM typewriter.*

ABCDEFGHIJKLMNOPQRSTUVWXYZ ?%#@±$

abcdefghijklmnopqrstuvwxyz & 1234567890

out letters on a plastic strip template. The template is placed against a T-square and, while the stylus traces a grooved letter, the pen produces an inked letter. This type of lettering, if well spaced, looks neat and is relatively fast to produce. The alphabets are limited in style but templates are available in various sizes with upper and lower-case, numerals, and special symbols.

Electronic typewriters offer a variety of fonts and typestyles. The copy can be quickly typed for paste-up. However, some typewritten lettering does not reproduce as well in print, with reduction, as does lettering prepared by other methods.

Laser Prints

Laser prints that operate in tandem with a graphics computer and software are gaining popularity as a quick method of producing copies of type, simple drawings, and textures. Cost determines the degree of quality; the more expensive the computer and its software, the smoother and more uniform the lines and image density of the laser print will be. Line art created on a low resolution computer and output for laser prints will probably be jagged (staircased). Blacks may not be solid and will contain spots or streaks of gray. Uncoated paper used for a laser print will produce a fuzzy appearance.

On the other hand, laser prints on coated paper, imaged by a high-resolution graphics computer, are generally acceptable for publication. Lines are smooth and clean, and blacks are uniformly black. Reducing the image photographically to perhaps a 5 × 7 inch print will further enhance the crispness of the image.

Letter Spacing

Letters should be spaced visually, taking into account the shapes of the letters and the open spaces around them. Letters that have been spaced mathematically have a choppy, unattractive appearance. A, C, L, and T require less space between them than closed letters such as H, M, and N. As a general rule, good letter spacing makes letters appear to occupy equal areas.

FIG. III–24: *Mathematical vs. visual spacing. A. Spaced mathematically. B. Spaced visually.*

A CAL I FORN I A

B CALIFORNIA

"Kerning" is the selective adjustment of letter spacing between certain combinations of letters, while the rest of the spacing in a line remains the same. Some combinations are visually improved by decreasing the space between them, such as Te, Ta, Ve, AW, YA, etc.

Word Spacing

The space between words should equal 3/4 of a lower-case m. Punctuation marks should not be given a full letter space. If words are too close they tend to run together, and if too far apart the white spaces tend to separate the text.

Leading

The distance between lines is called leading (pronounced "ledding") and is an important consideration when there are more than several lines of type. Leading is expressed in points. Generally, small type sizes require less leading, and larger type sizes more. When labeling a drawing or diagram, there should be a space of at least two lines above and below each label.

Labels and Leaders

Labels and leaders fall into the category of line art. Labels are words, letters, numbers, arrows, or symbols applied to illustrations. Captions, or legends, are not labels but rather text material that is typeset with the rest of the manuscript by the printer's typesetter. Labels can be prepared by hand lettering or by scribers (Leroy, Wrico, etc.), or with transfer type, cut-out alphabets, typed tapes from lettering machines, or by phototypesetting and paste-up.

Labels should not be placed too close to an illustration and should be arranged so that they lie within the vertical or horizontal format of the drawing. Labeling is an integral part of the overall composition of

FIG. III–25: *Examples of leading.*

8 point Times Modern

8/8 Set solid: no leading

Check to see that letters or paragraphs are not repeated or transposed. Defective characters occur less frequently with photocomposition than with hot metal composition; remember, photocopies may not be clear reproductions of the original impression. Copy editors at the printer or publisher are responsible for correcting poor typography and alignment, but authors should verify the proper alignment of statistical or chemical data and equations. Symbols should be checked for agreement with recognized conventions (*see* chapters 12 and 13).

8/10 2 point leading

Check to see that letters or paragraphs are not repeated or transposed. Defective characters occur less frequently with photocomposition than with hot metal composition; remember, photocopies may not be clear reproductions of the original impression. Copy editors at the printer or publisher are responsible for correcting poor typography and alignment, but authors should verify the proper alignment of statistical or chemical data and equations. Symbols should be checked for agreement with recognized conventions (*see* chapters 12 and 13).

8/12 4 point leading

Check to see that letters or paragraphs are not repeated or transposed. Defective characters occur less frequently with photocomposition than with hot metal composition; remember, photocopies may not be clear reproductions of the original impression. Copy editors at the printer or publisher are responsible for correcting poor typography and alignment, but authors should verify the proper alignment of statistical or chemical data and equations. Symbols should be checked for agreement with recognized conventions (*see* chapters 12 and 13).

an illustration and should not cover, cross over nor obscure any of the central focus of the artwork. Sometimes labels will fit within a structure, but clarity is lost if part of the label is too near an edge or is surrounded by lines leading to other structures. Labels should not be so large or numerous that they overwhelm the central message of an illustration and they should be readable with the page held in a normal position or turned 90 degrees clockwise.

When labels are applied to a group of illustrations that will appear in one article or one textbook, the point size of all the type or lettering should be the same in the final printed product. Therefore, if labels are applied by the artist, each illustration should be planned for uniform reduction of the lettering. If labels are applied by the printer, instructions should be written on a tissue overlay stating the desired placement

FIG. III–26: *A. Labels that are poorly placed and oversized distract attention from the drawing. B. Labels properly sized and placed inside and outside the drawing enhance reader's understanding. (Artist: Elaine R. S. Hodges. Reproduced, with permission, from Hodges, RW:* Agonopterix hesphoea, *male. J. Lepidopterists' Soc.)*

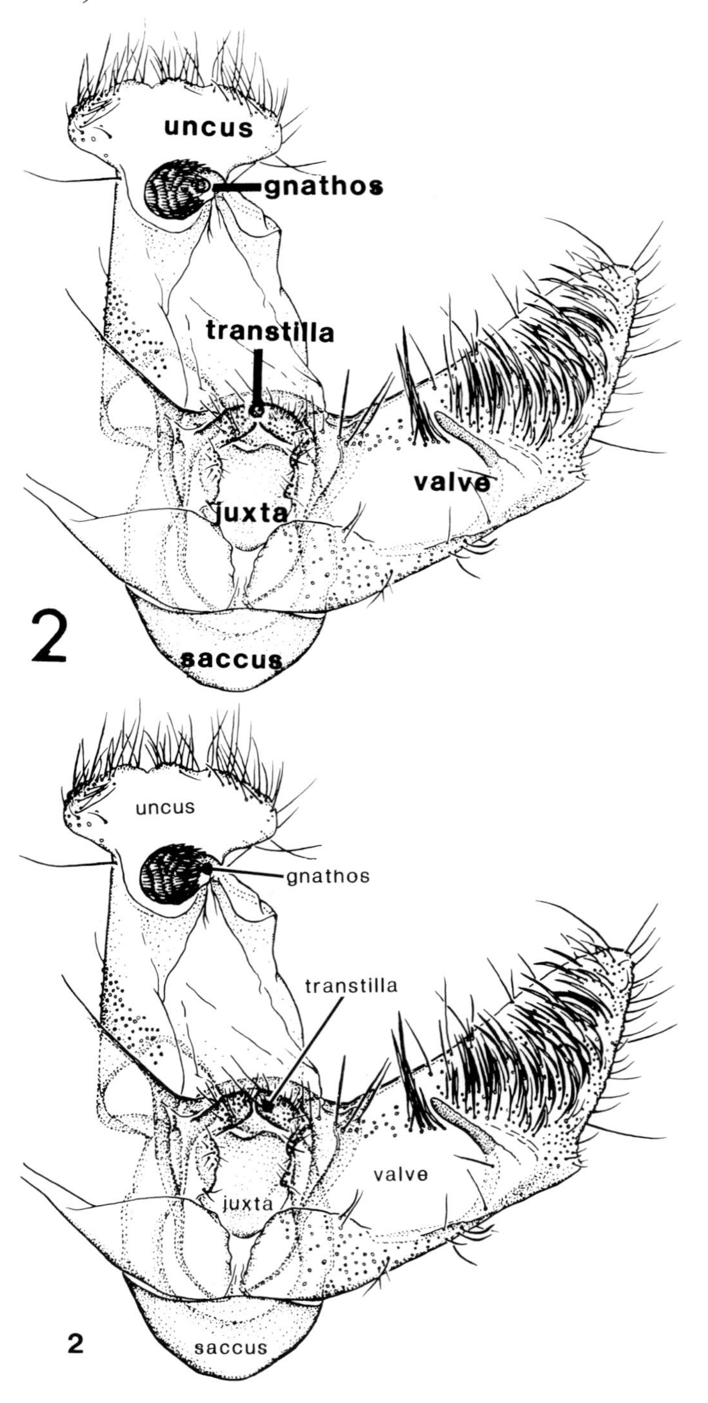

FIG. III–27: *Examples of various types of leaders.*

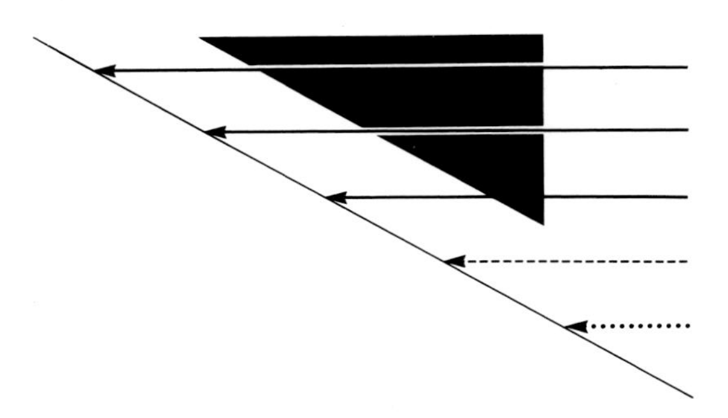

and point size. Type size for labels should never be smaller than 8 points (approximately 2 mm) in the final printed piece.

If the author or illustrator has an additional use for the illustration that requires other labels or another treatment, then the labels for the reproduction can be placed on a transparent, registered overlay. The overlay is photographed in place over line artwork, but photographed separately from color or tone artwork.

Leaders are lines that connect the label to its corresponding part of an illustration. They may be solid or may consist of rows of dashes or dots that guide the eye across the page and provide a visual connection between widely spaced materials.

It is important that leaders not detract from the drawing. Heavy leaders crossing a delicate drawing compromise the effect of the rendering. Leaders should be noticed after the artwork, and should be separated from each other as much as possible. They should not cross each other nor cross over a large area of an illustration unless absolutely necessary. If the leaders go to a specific area they may end with a dot or bullet, but generally not with an arrow, since arrows also are used to indicate direction.

Leaders from labels lying to the left of the drawing should begin 2/3 of the way up the last letter of the first line. Leaders from labels on the right of a drawing should begin 2/3 of the way up the first letter. Light and medium weight leaders work well in complex illustrations. Heavy or bold leaders and labels are more suitable for diagrams.

If many leaders are required in a drawing they can be deemphasized by employing dashed or dotted lines. It is important, however, that the space between the dashes or dots is not too great. When a leader crosses part of an illustration a thin line of artwork immediately next to the leader, toward the light source, should be whited out, especially when the leader passes over a dark area. Customarily, the imaginary light source in scientific illustrations comes from the top left corner of the drawing. Highlights on the bottom of leaders are misplaced.

Direction lines and symbols should be placed in open areas. Arrows,

FIG. III–28: *A. Leaders and labels positioned outside drawing. B. Leader from left label begins 2/3 of the way up the last letter on first line. C. Leader from right label begins 2/3 of the way up first letter on first line. (Artist: Nelva B. Richardson. Reproduced, with permission, from Tortora, G: Human Anatomy. San Francisco, CA, Canfield Press, 1977.)*

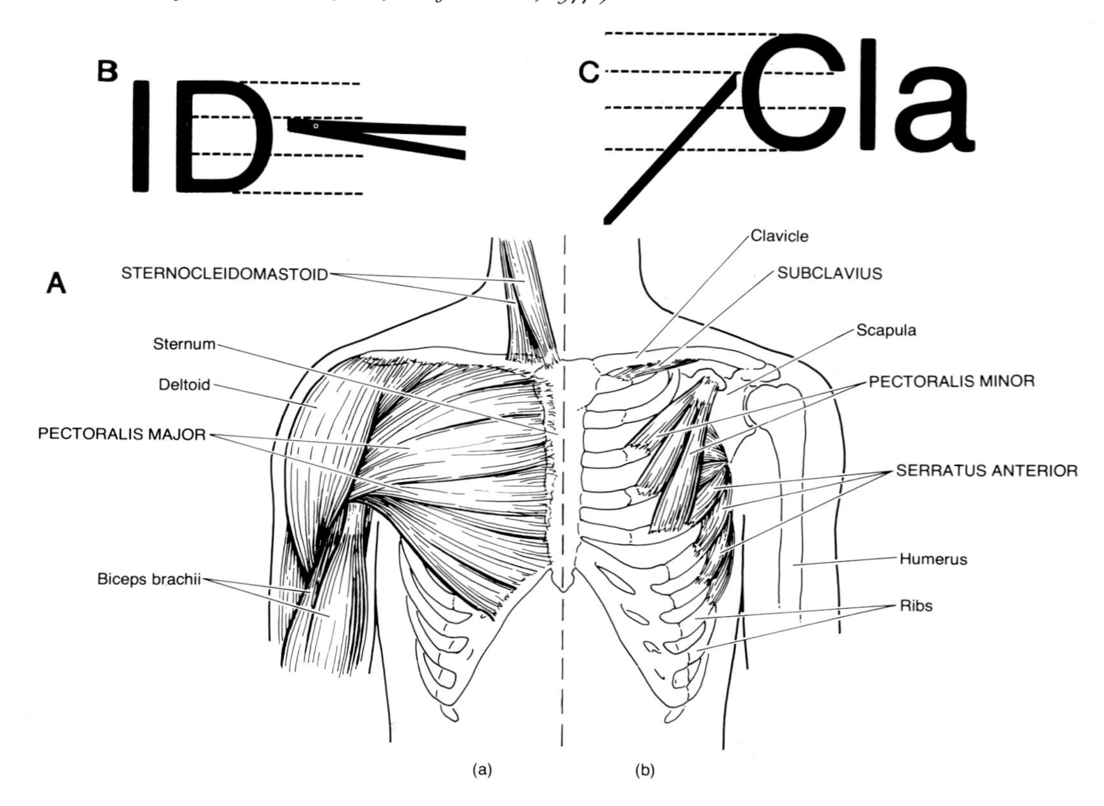

which are often very important in medical and scientific illustration, should be used sparingly but effectively. The weight of the arrow should be in direct proportion to the importance of the subject it points to or the direction it indicates, or to the relative impact of its thrust.

FIG. III–29: *Highlight on leader as it crosses the drawing is placed to reflect imaginary light source at top left.*

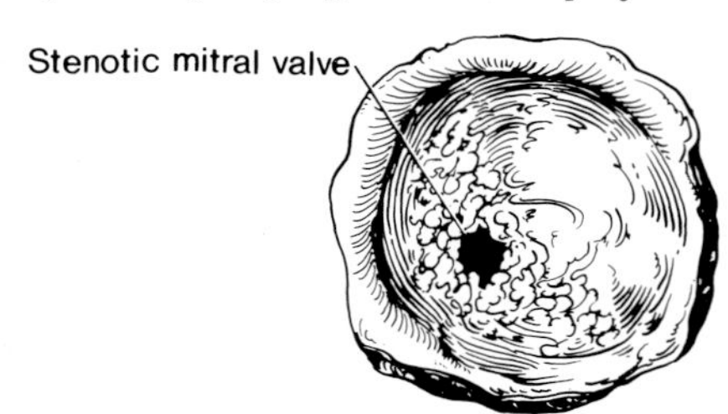

FIG. III–30: *The circle frames an insert that shows an enlarged view of the placement of sutures. (Artist: Nelva B. Richardson. Reproduced, with permission, from Donald, PJ: Head and neck cancer: management of difficult cases. Philadelphia, PA, W. B. Saunders, 1980.)*

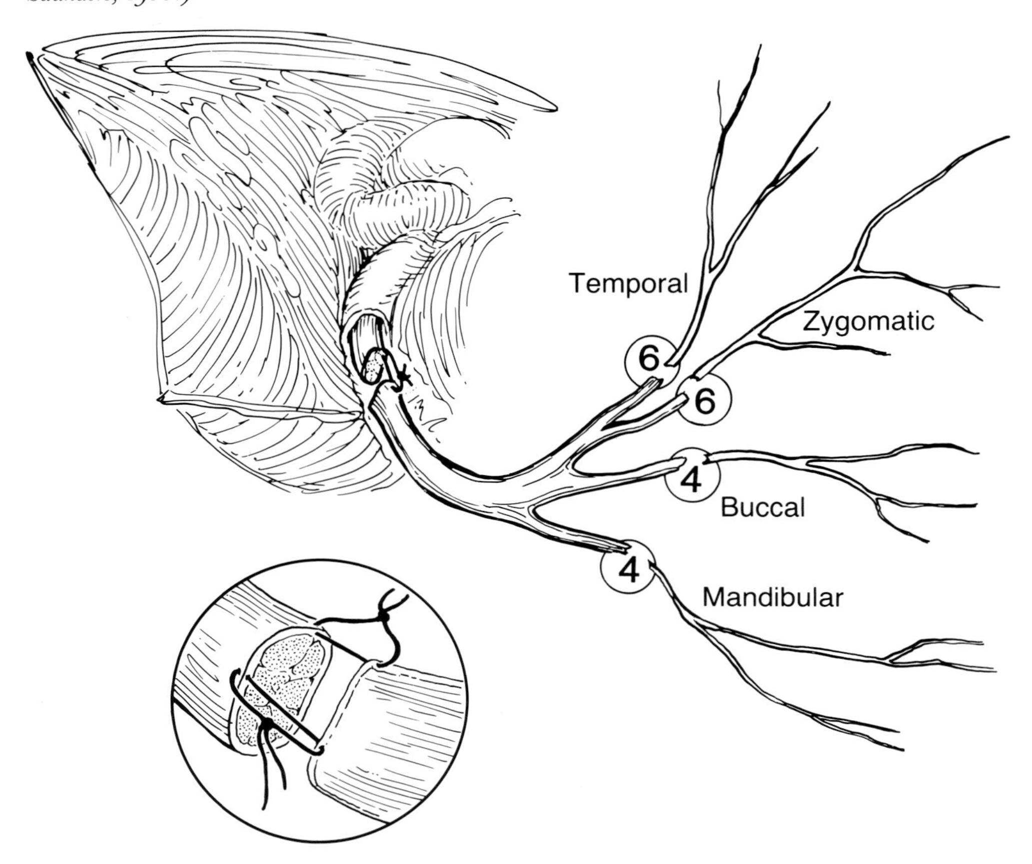

Reduction of Line Art

Planning for the ideal reduction of 67% to 50% of original size permits the artist to complete any illustration with lines and space between the lines that will reproduce satisfactorily. To determine how much reduction a stippled, contoured, or cross-hatched drawing can take, look at the space between the lines or dots. For 50% reduction, the space should be at least equal in size to its adjacent lines or dots. If the printer is expected to reproduce line art well and to maintain all details and quality, the rules of reduction must be observed by all concerned.

If lightweight, uncoated paper of inferior quality is to be used in reproduction, the artwork should not contain elaborate or fine-line rendering. Planning for less reduction from final art to printed page is preferable if it is known that such paper will be used in the publication.

Frames

Frames are used to separate different components on the same illustration. Care must be used in drawing frames, since a weak line is ineffective and one that is too dark or heavy is distracting.

Mechanical Drawings

Mechanical drawings are generally limited to the use of outline alone. They must give exact and positive information regarding details of a structure and its function. Rather than producing a free-hand drawing, the artist uses instruments such as triangles, T-squares, French curves, or templates to achieve a clean, accurate rendering. Each line should be uniform and may range in width from medium to bold in various areas of the drawing. The basic principles of good composition apply to mechanical drawings as well as to all other types of illustration.

Grouping Illustrations

Grouping related illustrations often enhances reader comprehension. When a series of illustrations is presented as a unit with a linking figure legend, the comparison is more quickly apparent than if the illustrations are scattered through the text. Grouping also saves space when the layout is planned to fit the number and width of the columns of type on a page.

Frequently, instructions for grouping illustrations are specified by the author in an accompanying sketch or diagram. When the illustrations

FIG. III–31: *Mechanical drawing of an endoscopic apparatus. (Artist: Laurel V. Schaubert. Reproduced with permission of the artist.)*

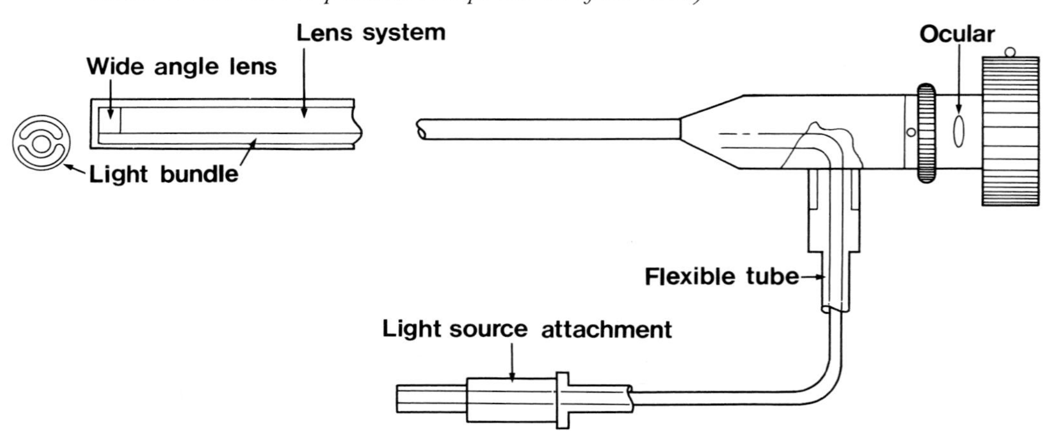

FIG. III–32: *Example of grouped illustrations. Embryological development of the human ear. (Artist: Nelva B. Richardson. Reproduced, with permission, from Donald, PJ: Head and neck cancer: management of difficult cases. Philadelphia, PA, W. B. Saunders, 1980.)*

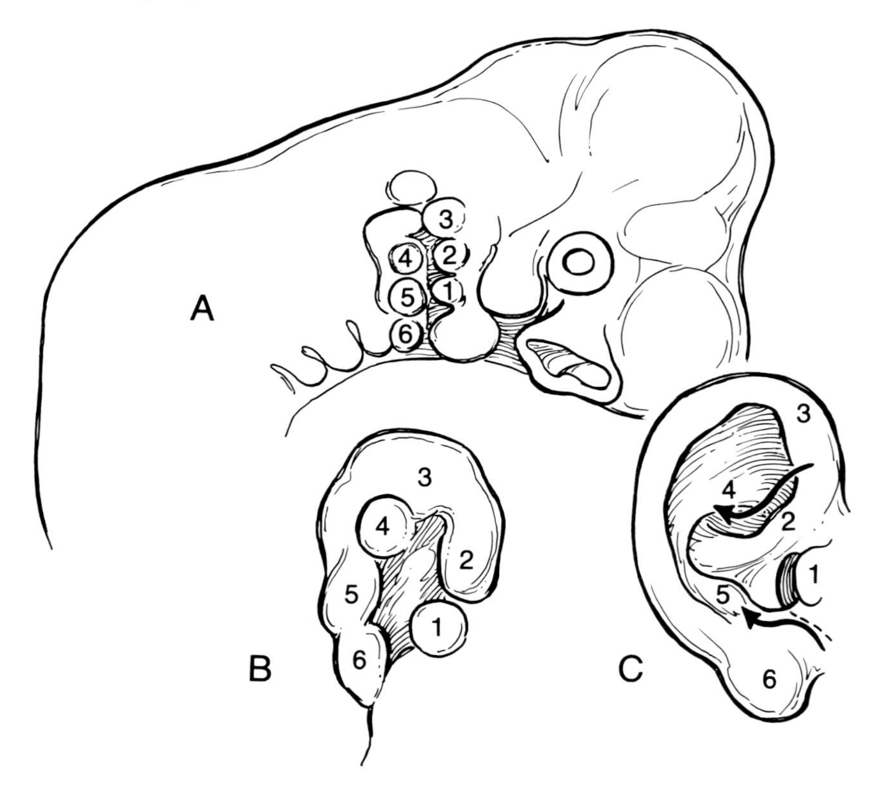

are prepared professionally, the artist will group the figures appropriately in the final rendering. In the absence of instructions from the author, the editor may wish to take the initiative if grouping would obviously improve the presentation.

Overlays

Purposes

An overlay is a translucent or transparent film or tissue, containing artwork or type, which is placed on top of an illustration or mounted photograph. Its purpose is to add information to the base art. It must be precisely aligned with the base art with register marks. These are placed on the base art and on each overlay in three or four corners. The register marks should lie in the outside margins and not impinge on the rendered area, but should be visible within the camera's frame image. Some register marks that are transferred from a transfer-type carrier sheet may discolor the artwork in time, and should be placed at least one inch outside the boundary of the drawing. Often, the best way to affix register marks is to draw them with ink.

An overlay is used for placement of labels and leader lines when more than one set of labels is to be used. If the base art is a continuous tone or color drawing, or a photograph, labels and leader lines should always be placed on a transparent overlay. Labels placed directly on continuous tone artwork will be reproduced with a halftone screen and will appear

FIG. III–33: *The preparation and use of "windows." A. Black windows are created on artwork. B. Negative film of artwork converts original black "windows" into clear "windows." C. Final printing: halftone negatives were "dropped" into the clear "windows," giving a printed image that combines line (letters) with tone (photographs).*

A

CPR: A TRAINING FILM
16 mm, COLOR, 33 MINUTES

PRICE LIST:	16 mm film	Video-cassette
Industry and government	$350	$275
Non-profit organization	$315	$250
American Red Cross	$280	$225
Replacement film	$175	$130
Rental	$35	—

If your copy of CPR: A TRAINING FILM is damaged or wears out due to continuous use, return it to us and we will send you a new film at ½ the price of the original. (The replacement price is the same for all groups.)

You will be notified of any changes in CPR techniques that may be recommended by the American Red Cross and of the cost of updating. You may send us the film for splicing or may request only the updated segment, which we will send to you promptly.

Write or call: THANCO FILMS, 3047 Baker Street, San Francisco, CA 94123, (415) 921-0151.

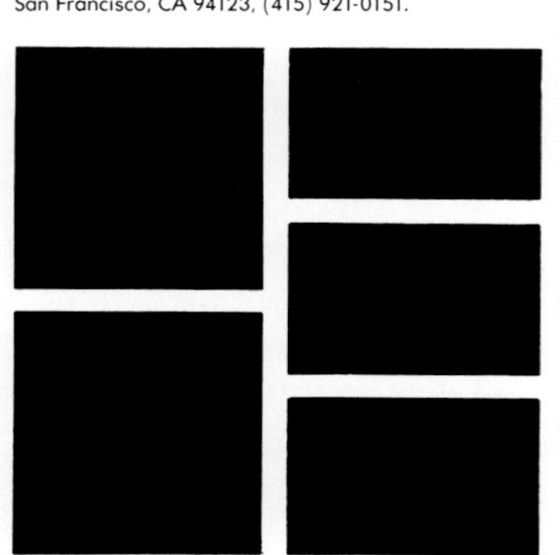

B

CPR: A TRAINING FILM
16 mm, COLOR, 33 MINUTES

PRICE LIST:	16 mm film	Video-cassette
Industry and government	$350	$275
Non-profit organization	$315	$250
American Red Cross	$280	$225
Replacement film	$175	$130
Rental	$35	—

If your copy of CPR: A TRAINING FILM is damaged or wears out due to continuous use, return it to us and we will send you a new film at ½ the price of the original. (The replacement price is the same for all groups.)

You will be notified of any changes in CPR techniques that may be recommended by the American Red Cross and of the cost of updating. You may send us the film for splicing or may request only the updated segment, which we will send to you promptly.

Write or call: THANCO FILMS, 3047 Baker Street, San Francisco, CA 94123, (415) 921-0151.

C

CPR: A TRAINING FILM
16 mm, COLOR, 33 MINUTES

PRICE LIST:	16 mm film	Video-cassette
Industry and government	$350	$275
Non-profit organization	$315	$250
American Red Cross	$280	$225
Replacement film	$175	$130
Rental	$35	—

If your copy of CPR: A TRAINING FILM is damaged or wears out due to continuous use, return it to us and we will send you a new film at ½ the price of the original. (The replacement price is the same for all groups.)

You will be notified of any changes in CPR techniques that may be recommended by the American Red Cross and of the cost of updating. You may send us the film for splicing or may request only the updated segment, which we will send to you promptly.

Write or call: THANCO FILMS, 3047 Baker Street, San Francisco, CA 94123, (415) 921-0151.

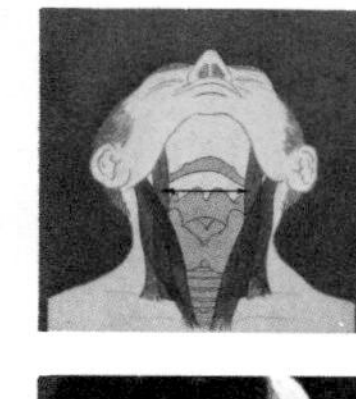

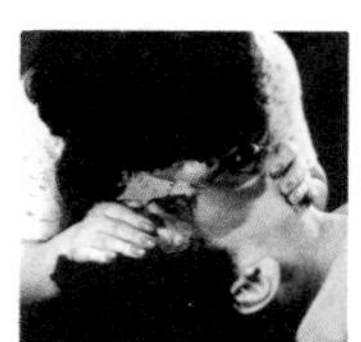

gray rather than black in the final printed piece. In certain instances, line art is prepared on a registered overlay that will be printed over continuous tone or color art. Camera-ready labels can be affixed to the overlay, or the artist can provide a tissue overlay with handwritten instructions to the printer showing precisely where the labels are to be placed.

Overlays are also used to create flat tones or flat color that will enhance the central message. Each tone value is prepared on a separate, registered overlay, and the percentage of 100% density (for example, 20%, 40%, etc.) is written on it as an instruction to the printer. When photographed, black areas over the drawing where a specified screen will be used show up as clear "windows" on the printer's negative. Color or a screen is then printed in the "window" in register with the base drawing. The artist may create these black areas with ink or paint on a translucent film or vellum, or may use Amberlith or Rubylith. These are clear polyester sheets with an adherent orange or red film which is also transparent but which will appear black to the camera. The areas that will be black are easily cut out and the unwanted pieces of film peeled away. The result is a clear acetate overlay with an orange or red film covering those areas of the drawing that will receive color or a screen. A wax-backed plastic film that is more opaque, dense, and usually deep red is also available. It can be cut out, laid down on vellum or film, and trimmed to produce a clean-edged black area.

Printer's Instructions and Process

Along with register marks and identification it is important that the mounts for the base art and each overlay have their own instructions to the printer written in the lower right corner, well away from the boundary of the artwork. These instructions must include the percentage of reduction, which will be the same for the base art and all overlays; the percentage of tone value, which will be 100% black for the base art; and selected percentages for each overlay and the preferred line size of the screen. The printer will then be able to reproduce the art according to the artist's instructions.

The base art is photographed by the printer without a screen, 100% black and reduced as directed. A printing negative is made of the exact size that the work will be when printed. Each overlay is then photographed separately with the same reduction and according to the specifications for percentage of tone and screen, thus creating clear areas of the blacked-out areas. Precise alignment of the registration marks on all the negatives enables a composite film negative to be made for the press run.

FIG. III–34: *Instructions to the printer. The base art shows identification of the figure, registration and crop marks, reduction, and percentage of black. The overlay shows the area to be printed in color or screen, the PMS number for the color, and the percentage of the screen. Register marks on the overlay are placed to exactly match those on the base art.*

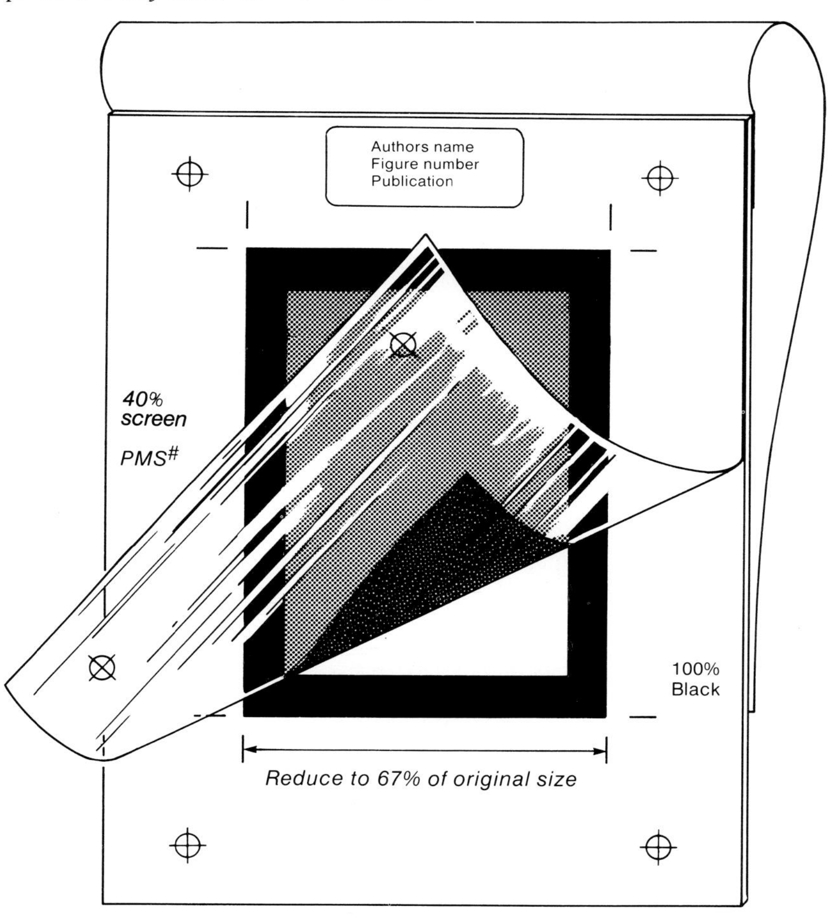

Markers

Markers appearing on an illustration routinely include the author's orientation of the figure and instructions from the artist/author/editor to the printer to ensure appropriate reproduction.

A label clearly noting the "top" should be placed on the top front of the mount. The artist or author may also wish to place crop marks on certain illustrations to eliminate areas that do not contribute to overall effectiveness. These crop marks indicate where the figure can be successfully limited by masking. Crop marks should be placed on the margin of the mounting board or on a tracing paper overlay, but never directly on the artwork or photograph.

Publication Standards

Original Art

Each generation of reproduction loses some of the detail and contrast found in the original. Therefore, to ensure optimal results on the printed page only original artwork should be submitted to the printer for reproduction. This is especially critical with continuous tone and color illustrations.

If the original work absolutely is not available, then every effort should be made to provide the printer with a photographic print of the highest quality possible or with a negative of the original. Xerographic copies are useful for editorial review of a manuscript, but can com-

FIG. III–35: *A. First: reproduction from original art. B. Second: reproduction from first reproduction. C. Third: reproduction from printed page. (Artist: Laurel V. Schaubert. Reproduced with permission from the publisher. Way, LW (ed.): Current surgical diagnosis and treatment. Los Altos, CA, Lange Medical Publications, 6 ed., 1983.)*

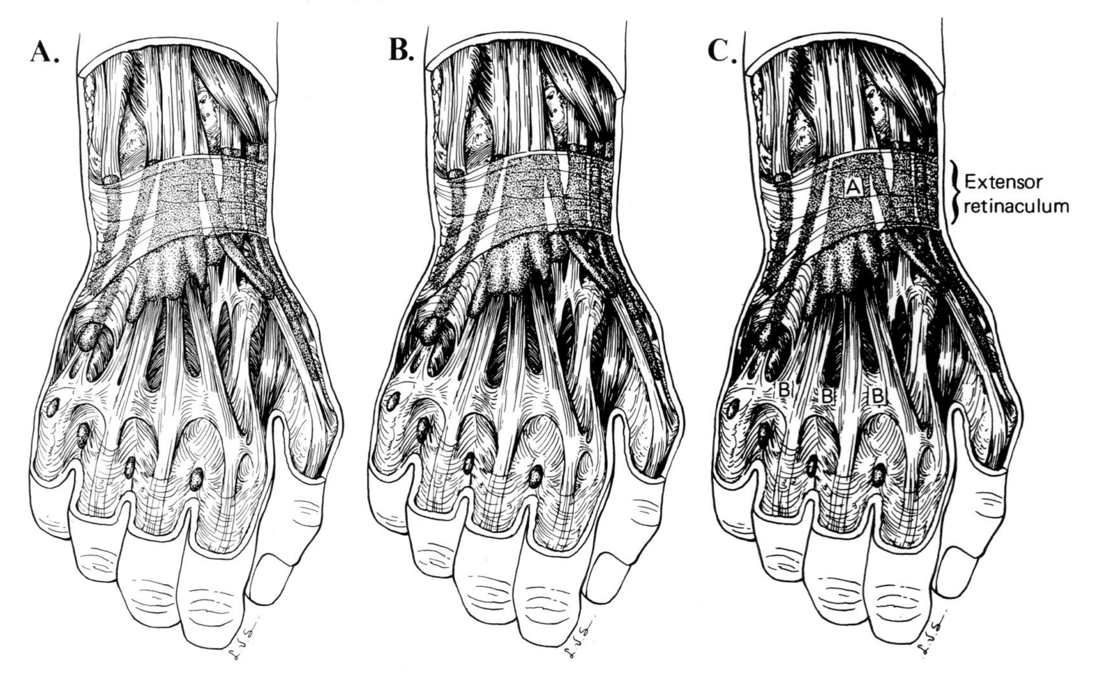

promise the end product if they are reproduced in print. They should be marked on the front, "not for reproduction."

Changes and Corrections

Once the artwork has reached the printer, it should already have been reviewed by artist and author, and corrections generally are not required. If for any reason changes must be made, or if the artwork is damaged in handling or production, the corrections should be made by the original artist or someone the artist designates. Opaquing negatives of artwork to eliminate extraneous spots or artifacts is usually done by the printer.

Tone Art

Advantages

If a drawing is expected to depict a realistic view of a subject and the cost of printing in color is beyond the production budget, then a continuous tone rendering in black and white is the ideal choice. This type of illustration, created by an experienced artist, can portray the full range of gray values from black to white. The technique is especially valuable in surgical or anatomic drawings, since it permits clear differentiation among different kinds of tissue. A tone drawing can so closely match the realism of a photograph that in print it is sometimes mistaken for one. This realism has the effect of assuring the viewer that the illustrated information is accurate, since it portrays "life" as one is accustomed to seeing it.

The question is frequently asked, "Why not use a photograph of the subject, rather than a drawing?" The answer is simply that the camera does not have a mind; it can only document what it sees. It cannot make decisions to eliminate unwanted detail, change the perspective for a better view, show relationships of subsurface structures, or portray a series of events in a single illustration. The mind of the artist, however, can conceive a variety of illustrative approaches and treatments which can provide instructive, understandable, and pleasing results.

Techniques

Tone drawings can be produced with any medium that permits the artist to render gradations from black to white. These include graphite, charcoal, watercolor, gouache, acrylics, and tempera. The medium can be

FIG. III–36: *Continuous tone illustration. Note realism of tissues and effect of semitransparency in fluid-filled ovarian cyst. (Artist: Ralph Sweet.)*

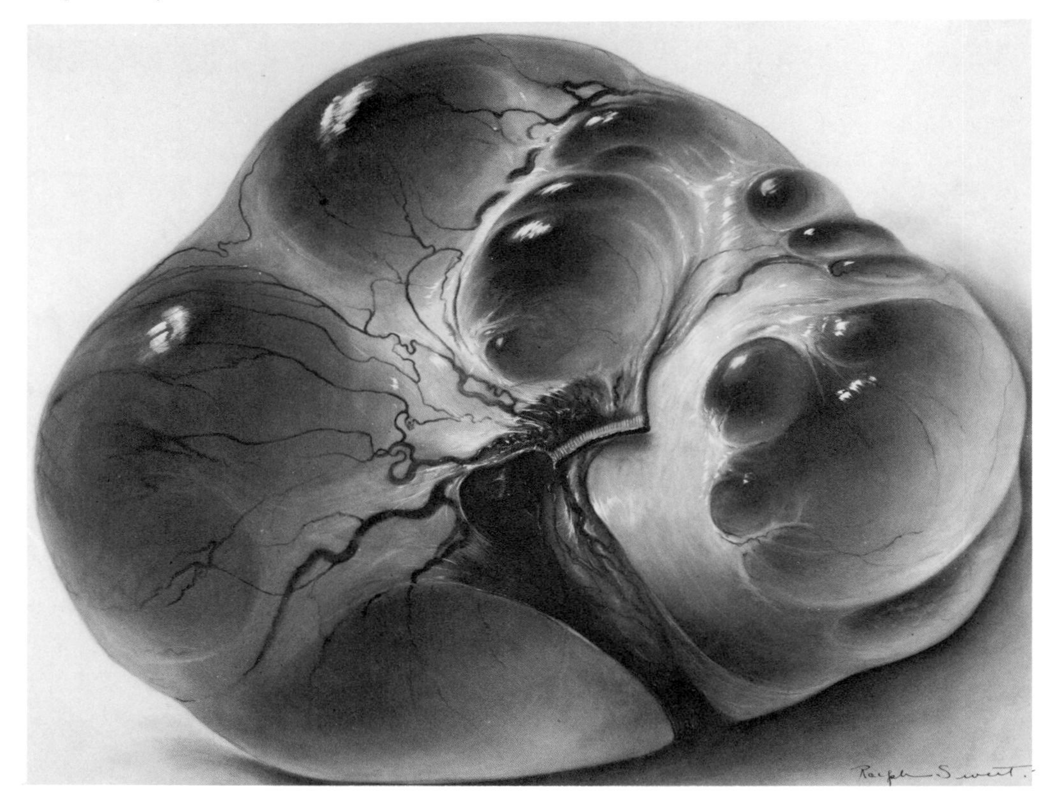

applied with a pencil, brush, or airbrush. Generally, a good grade of illustration board is used as a ground; if it is reasonably stiff, it will not require mounting. Some varieties of chalk-embedded papers are likely to crack with handling and should always be mounted on a rigid backing before the artwork is begun. Tone drawings prepared on matte film require a white board as backing. Various fixatives are available to "fix" the drawing's surface. However, some brands may change the character of the artwork and should therefore be applied only by the artist.

Carbon Dust

This classic method of rendering tone was originally developed and refined at the first school of medical illustration in the United States, The Johns Hopkins University School of Medicine, by the late, great Professor Max Brödel. The method is sometimes referred to as the "Brödel Technique" and is still used by many illustrators, although the original chalk-embedded paper with a fine stipple surface (Ross board), on which the drawings were rendered, is no longer manufactured. How-

FIG. III–37: *Continuous tone on Ross board. (Artist: Max Brödel. Reproduced with permission from the Johns Hopkins University School of Medicine.)*

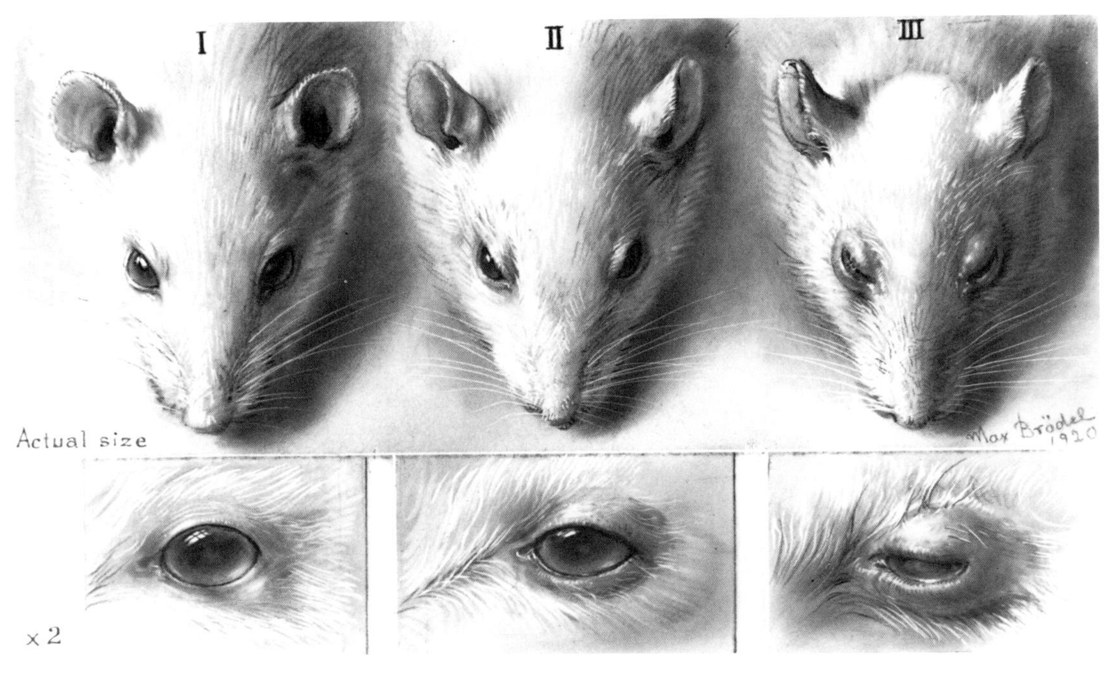

ever, other similar grounds are available today, and the technique can also be successfully employed on a matte finish polyester or acetate. A sketch is made with a carbon pencil, transferred to the ground, and worked up with sable brushes dipped in finely pulverized carbon dust (sometimes called "sauce"). Additional drawing with carbon pencil is done for emphasis or detail; middle values and highlights are developed with chamois, erasers, cork stomps, or knife scratches. The result is a rendering that is lifelike, with realistic detail, texture, and lighting.

Wash

Monochromatic watercolor washes provide a fast and effective technique for developing tone illustrations. Transparent blackish and brownish tints are mixed with water to the desired intensity and applied with watercolor brushes to a fine grade of illustration board or watercolor paper. Dark detail is developed with fine brush strokes, and highlights appear either as spots of white paper left unpainted, or are applied with opaque white gouache.

Skill with the general principles and techniques of watercolor painting is essential to the success of this type of rendering. If well done, however, wash drawings are realistic and reproduce well in print.

Monochromatic, transparent washes also can be applied with an air-

FIG. III–38: *Continuous tone drawing on Cronaflex. This material is translucent and can be taped over the original sketch and rendered without making a transfer.* Chloronia heiroglyphica. *Neuroptera: larval head, dorsal and ventral. (Artist: Elaine R. S. Hodges. Reproduced with permission. Penny ND, Flint, Jr., OS: A revision of the genus* Chloronia. *Smithsonian Contributions to Zoology. Washington, DC, The Smithsonian Institution, 1982.)*

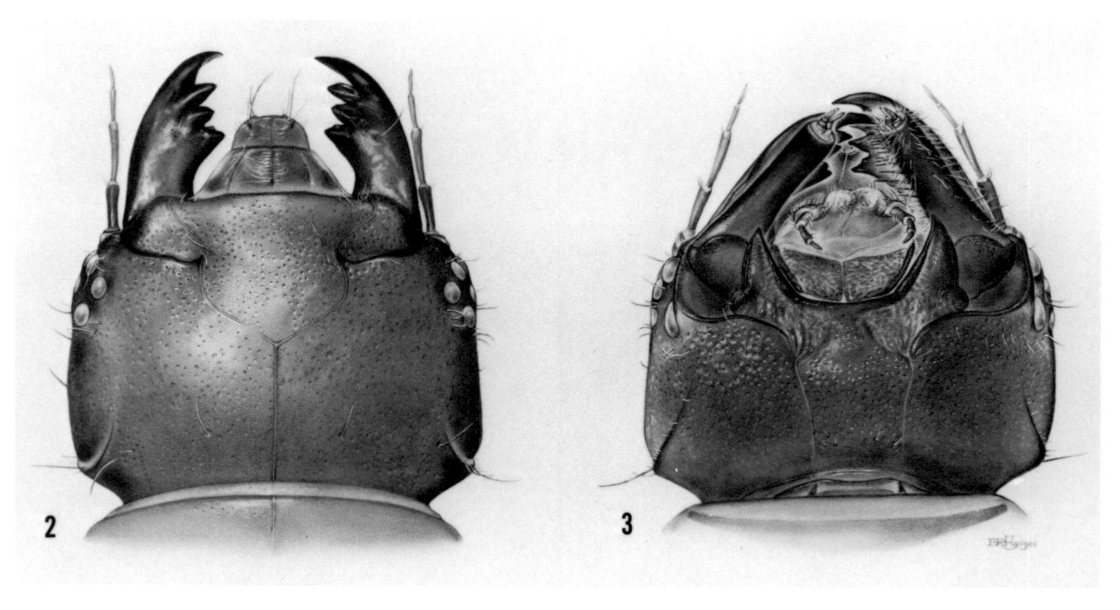

FIG. III–39: *Line art can also be quickly and effectively rendered on film taped over a pencil sketch. (Artist: Jane Hurd. Reproduced with permission of the artist.)*

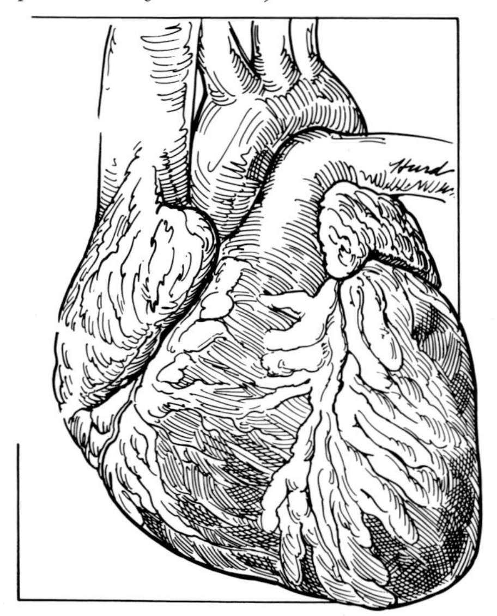

FIG. III–40: *Continuous tone illustration created with monochromatic watercolor washes. (Artist: Tom Jones. Reproduced with permission from the Archive of Medical Visual Resources, The Francis A. Countway Library of Medicine, Boston, MA.)*

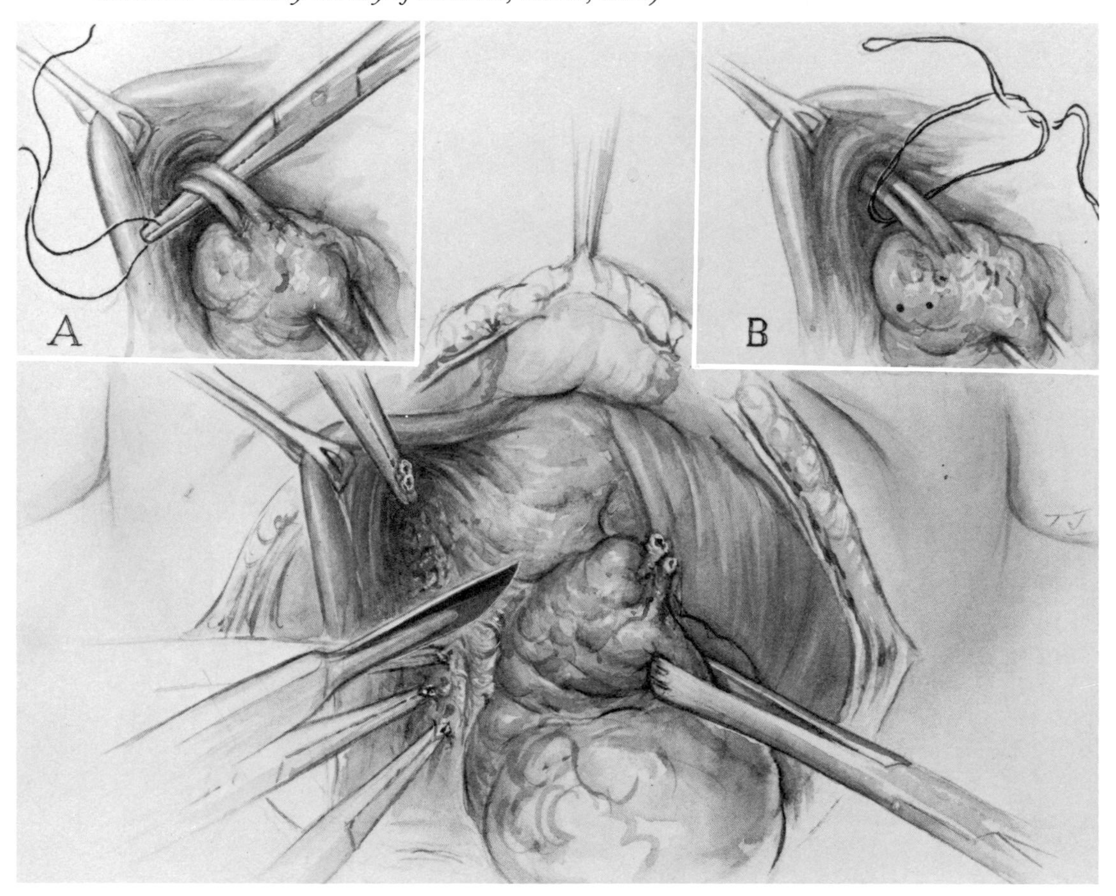

brush, which gives a less "painterly" effect but is soft and subtle. Detail can be added with drybrush, wash, or opaque gouache.

Graphite

Tone drawings are sometimes worked up entirely with graphite pencils. This technique enables the illustrator to carefully control values, texture, and detail, and in the hands of an expert the rendering is fast and effective. Leads of various degrees of hardness are usually employed to allow a gradual build-up of tones. A fine drawing paper mounted on a rigid backing, or a high grade of Bristol board, are the grounds of choice.

FIG. III–41: *Continuous tone illustration rendered with airbrush. (Artist: Robert J. Demarest. Reproduced with permission of the artist.)*

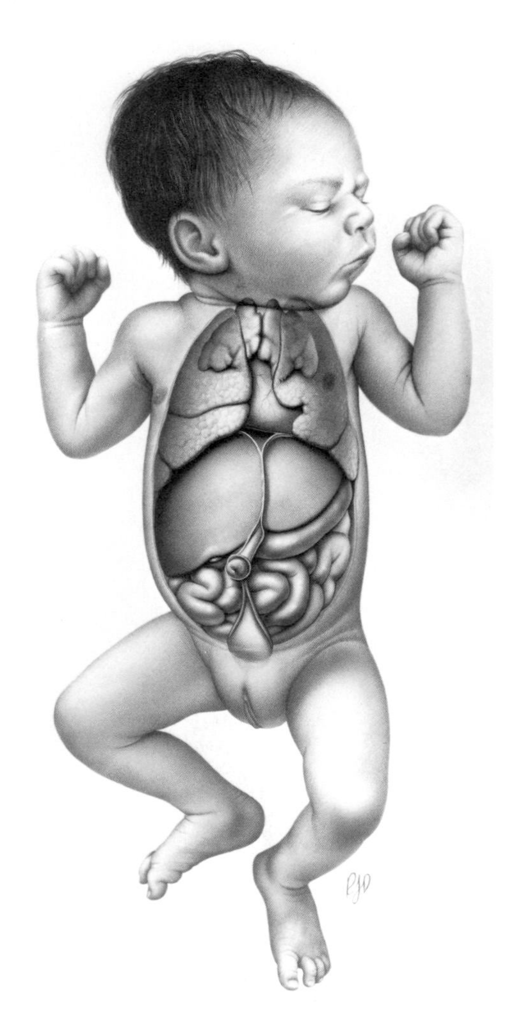

Labels on Tone Art

Since tone art is converted to a dot screen by the printer (See Chapter VIII, "Continuous Tone Photographs and Halftone Printing"), black lettering affixed to the face of the original artwork appears gray in print. To avoid this effect, labels and leaders can be put on a clear acetate overlay, in register with the drawing. Many publishers prefer to set the labels in type and apply them in their own shop. In this case, the artist indicates on a tissue overlay the exact positioning of the labels and leaders. Care must be taken in either event to consider the overall design

FIG. III–42: *Continuous tone drawing rendered with graphite pencils. Note the variations in values, created with pencils of various degrees of hardness. (Artist: Susan Strawg. Reproduced with permission. From "Habitat suitability index models: northern Bobwhite." U.S. Fish and Wildlife Service Biological Reports, no. 82, 1985.)*

and balance of the entire piece, and the placement and point size of labels should be planned along with the development of the artwork.

Double Image Prints

The problem of combining tone and line in one illustration often arises as, for example, with the need to add molecular weights and labels to a print of an electrophoretic gel. As a rule, the gel is photographed and a tone print is made. The artist then precisely cuts out those electrophoretic lanes that are to be used and carefully aligns and mounts them on a clean, white, opaque ground. The labels are then affixed to the ground adjacent to the lanes and the entire plate is photographed.

FIG. III–43: *A. Tone and line combined in a print produced from a single negative shot with tone film. Note the gray background and shadows at the edges of the pasted-up type and gel. B. Same image produced from two negatives. Tone film was used for the gel and line film for the type. Shadows are eliminated and the background is pure white.*

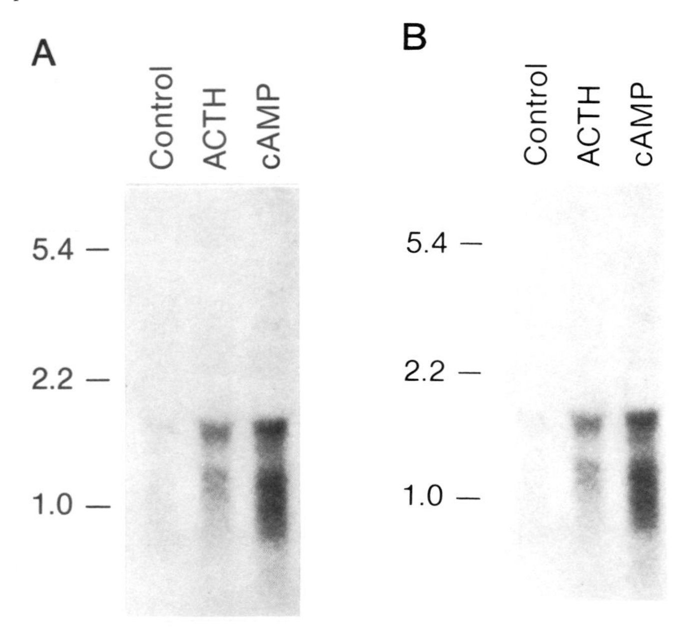

If the plate is photographed with tone film, the final print appears gray and shadows surround the pasted up lanes and typeset labels. This result is generally not acceptable for publication, but it can be avoided by making two negatives.

One negative is made of the lanes only, using tone film, with the labels and white ground masked out. The second negative is made with line film, masking out the tone lanes. The two negatives are then printed in register with each other on one piece of photographic paper. The final result displays a clean white background with sharp, black lettering, the tone strips in proper contrast, and no shadows around any of the elements.

Because this technique requires the additional steps of masking and shooting twice, the final print is somewhat more expensive to produce. However, the optimal results should be the primary concern.

Reproduction of Tone Art

To ensure the best possible results on the printed page, only original tone illustrations should be used for reproduction. The printer cannot restore original detail or clarity that has been lost in a photograph of a drawing. It behooves the illustrator to keep the size of the original art within manageable limits, and working 50% to 100% larger than the final printed size will ensure ease of handling in the printing process. However, convenience of handling can never outweigh the importance of final high quality in print. The only excuse for using a photograph of a drawing in reproduction is the total unavailability of the original work. If this is the case, then the printer should be provided with a high-quality photograph or, as a last resort, a negative.

Special care must be given to handling tone art. Even if a fixative has been applied to the surface, grease, food, water, saliva, or rough surfaces can mar the drawing, and only the skill of the original artist can repair damage done by careless handling. Sometimes the damage may be irreparable and a new drawing must be made.

Color Art

Color art prepared for publication in scientific journals must be planned in proportion to the column or printed page width, as with other types of artwork. It can be rendered in one or two colors or in full color, according to the preferences of the journal. It should be noted, however, that the cost of color printing can be significantly greater than printing line or continuous tone art, and the decision to submit color art for publication should be carefully weighed. Some publications ask the author to pay all or part of the costs of color printing.

Journals often request submission of color prints or transparencies, color negatives, or slides with a manuscript. Although the original artwork is always preferred for final production, prints or slides are sufficient for review. The original artwork is reserved for use by the printer only.

Preparing Art in Two Colors (Spot Color)

When it is desirable to enhance line art, charts, graphs, tables or even continuous tone illustrations, another color in addition to black may be considered. This can serve several purposes: to focus on an important area within a drawing; to subdue a bold line technique; to serve as a background screen and give the effect of pulling together widely distributed compositional elements; to add emphasis to bar and pie graphs; to distinguish graph lines from ordinate and abscissa lines; to add mass to outlined areas; or to create a mood.

Areas that require flat tones, either solid color or percentages of solid color (screens), are prepared with black or colored masking films (as de-

FIG. III–44: *Line art with two color screens applied. (Artist: Jane Hurd. Reproduced with permission of the artist.)*

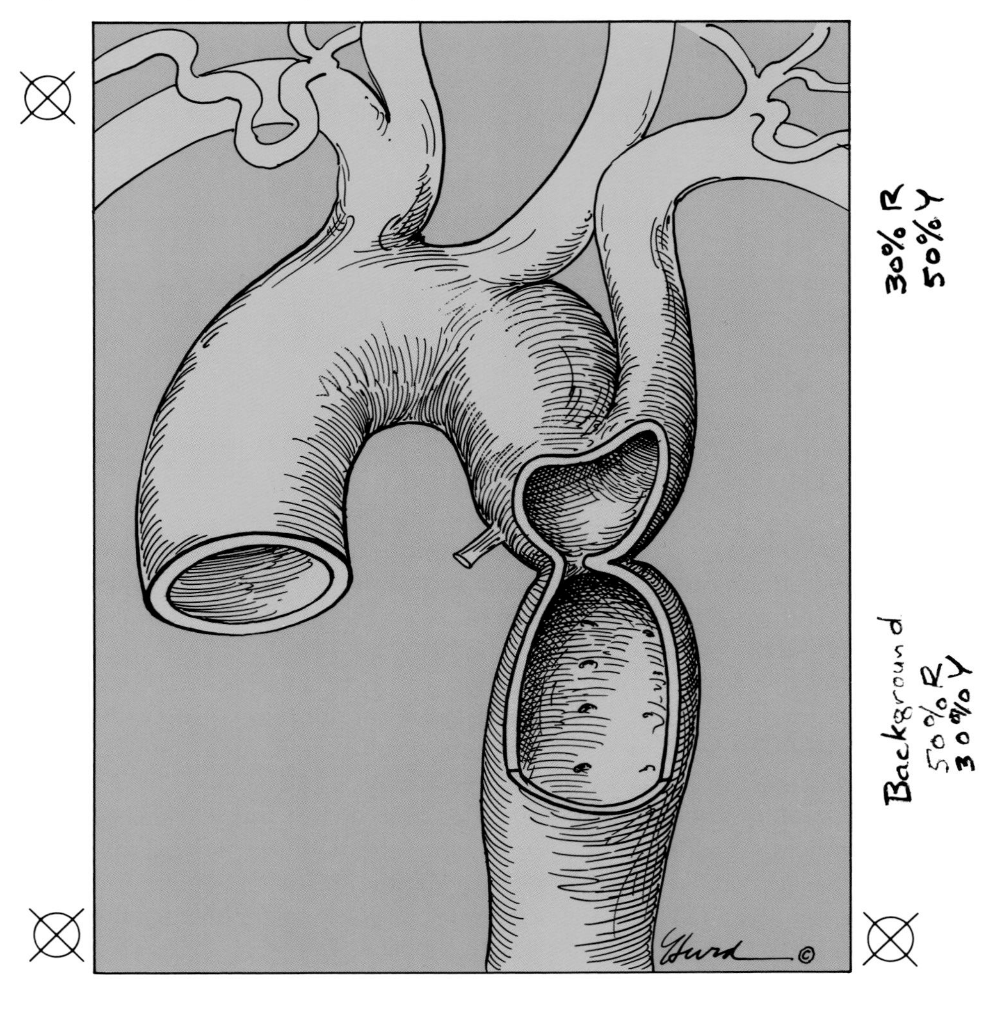

scribed earlier) or with keylines on a registered overlay. A keyline consists of thin black lines drawn with India ink which define the area that will receive the color. No actual color is applied by the artist. Instructions to the printer on a tissue flap note what process ink is to be used and what percentage of color is wanted. Solid or screened colors thereby appear uniform in print. It is important that the register of abutting colors is perfect, with no overlapping or uneven joining. Indications for the second color should always be placed on an overlay, whether using the keyline method, masking films, or painting the second color directly on a frosted acetate overlay. If a second color is made part of the base art, costly color separation must be done by the printer, a needless expense.

Rather than guessing at screen percentages, the illustrator should have in hand a printed color chart showing examples of printed screens of solid process inks from which an exact selection can be made. The standard reference is PMS (the Pantone Matching System), which can be obtained from most art supply stores and printers. Of course, all overlays should be in register with the base art. For other color-matching systems a printer can be consulted.

Preparing Full-color Art

Many journals do not print full-color illustrations in the manuscript section. There are certain instances, however, in which only color rendering can tell the story adequately. In these cases, the illustrator should be acquainted with the capabilities and limitations of the color process printing method, although it is not necessary to become an expert in the chemistry of printing inks nor in all phases of the printing process (see Chapter VIII).

Of primary concern to the illustrator is the selection of pigments from which a painting is created. Selection is determined by reproducibility in print. Several watercolor pigments that are almost impossible to reproduce with standard process inks are Payne's gray, ultramarine, cobalt, turquoise, Naples yellow, Indian yellow, cadmium yellow dark, yellow ochre, sepia, raw sienna, rose madder, thalo crimson, crimson red, Van Dyke brown, and burnt umber. These pigments are used by many illustrators, but fidelity in print is only approximate and often unsatisfactory.

Even black and white, when used indiscriminately, can cause printing problems. Permanent, Chinese, and bleed proof whites should be reserved for minute highlights and when great opacity of white is desired. When mixed with other colors, tints may appear true to the eye but not to the camera. Zinc white is preferable for achieving tints. Ivory black is preferable to lamp black. Muted pigments are best obtained when a color is mixed with its complement rather than with black.

Illustrators will have individual preferences, but most will agree that the following watercolor palette is adequate to render even the most detailed color illustration and to obtain true reproduction:

Primary palette:
- Zinc white
- Ivory black or Talen's photo-retouch in blackish and brownish
- Grumbacher's red and Winsor Newton's alizaron crimson (equal mixtures)
- Grumbacher's cadmium yellow, pale
- Winsor Newton's peacock or cerulean blue

Secondary palette:
- Winsor Newton's burnt sienna
- Winsor Newton's brilliant green or Grumbacher's permanent green, light
- Grumbacher's gamma neutral retouch grays (1–6)
- Bleed proof white (for highlights and opaquing)

Graphite or carbon pencil rubbed over colored pigment areas with stomps, dry brush, or corks may appear to give the effect of darkening or shading, but these substances have a coarse mineral content that can cause considerable problems for the engraver and ink color specialists.

It must be kept in mind that paintings prepared for the reproduction process require only a palette that will give the best fidelity when converted into printing inks.

The Preliminary Sketch

The underlying drawing, the plan, the structure on which color pigments will be placed basically determines the ultimate success or failure of the painting. For this reason, a preliminary sketch is made of the subject with graphite pencils, carbon or colored pencils, pastels, or thin washes. The composition of elements, perspective, focal points of emphasis, and highlight areas must be carefully worked out as the sketch develops. The direction and arrangement of leaders and the space for labels are all part of the composition. All of the thinking and planning should be completed before the rendering is begun.

The basic lines of the preliminary sketch that are essential guidelines are traced onto a tracing tissue. The back of this tissue is blackened with a #3 graphite pencil and excess graphite is removed with a soft tissue or utility brush. The essential lines are retraced with a hard, sharp pencil, which keeps the transferred lines thin and clean on the illustration board.

The Finished Artwork

Transparent watercolor

Using the preliminary sketch as a guide, the illustrator selects pigments that will give the desired result and a paper surface that best accepts the medium selected. Transparent watercolors or aniline dyes are suitable for non-textured, highly reflective structures, such as cysts, aqueous and vitreous humors, amnion and chorion, or topographic anatomy in which structures are shown in juxtaposition, or transparent forms containing subsurface structures. Hot-press illustration board is an ideal ground for such work.

FIG. III–45: *Full color illustration rendered with transparent watercolor. Chimpanzee portrait for the Primate House, Washington Park Zoo, Portland, Oregon. (Artist: Joel Ito. Reproduced with permission of the artist.)*

Gouache

Gouache pigments contain an opaquing substance that does not allow the white of the paper to show through. They are ideal for painting more solid, non-transparent structures such as bone, organs, and muscle. This medium does not blend easily or naturally, and several thin glazes are often necessary to achieve a desired effect. The effect of gouache can also be obtained by mixing transparent watercolor or dye with gouache zinc white or light (#1) neutral gray.

Acrylics

Acrylics are polymer emulsions in a resin vehicle. They can be diluted with water, but are water resistant once dry. In thin glazes they can be used similarly to the way transparent or opaque mediums are applied. Thin solutions of acrylics can be sprayed from an airbrush, painted in flat colors, or mixed with other transparent watercolors or dyes. A retardant can be added to prolong drying time. Acrylics can be painted over when thoroughly dry and will show no overlapping effect of underlying brush strokes nor become muddy. Paint brushes used for acrylics should always be washed with a gentle detergent, since acrylic allowed to dry on the brush becomes difficult or impossible to remove.

Dyes

Dyes, such as Dr. Martin's, are intense aniline solutions in pure liquid form with no vehicle. The colors are soluble and can be lightened by the addition of water. Because of their low viscosity, dyes are ideal for use in ruling pens to produce thin, solid lines in color. They also can be used as flat washes on a variety of grounds, such as illustration board, clay-surfaced boards, watercolor paper, layout tissue, and photographic surfaces. It should be noted that dyes fade easily, and artwork that is rendered with them should be promptly reproduced.

Tempera and casein

Tempera contains an albuminous or colloidal medium that often separates from the pigment and is difficult to remix. It is an opaque medium that can be combined with watercolor, thus enabling the illustrator to lay in dark base color transparently and then to work up the lighter values with opaques. It provides extensive control and facilitates correction.

Caseins are pigments that contain phosphoprotein of milk and lactic acid. If these colors are to be mixed with water and kept for any length of time, it is necessary to use distilled water or water that has been boiled to avoid the growth of bacteria. A ratio of 5 parts water to 1 part casein emulsion makes a strong, rich painting medium that can be applied with watercolor brushes or sable oil brushes. If tempera or casein (or gou-

FIG. III-46: *Full color illustration rendered with airbrush and acrylics. Bronchitis. (Artist: Ellen Going Jacobs. Reproduced with permission of the artist.)*

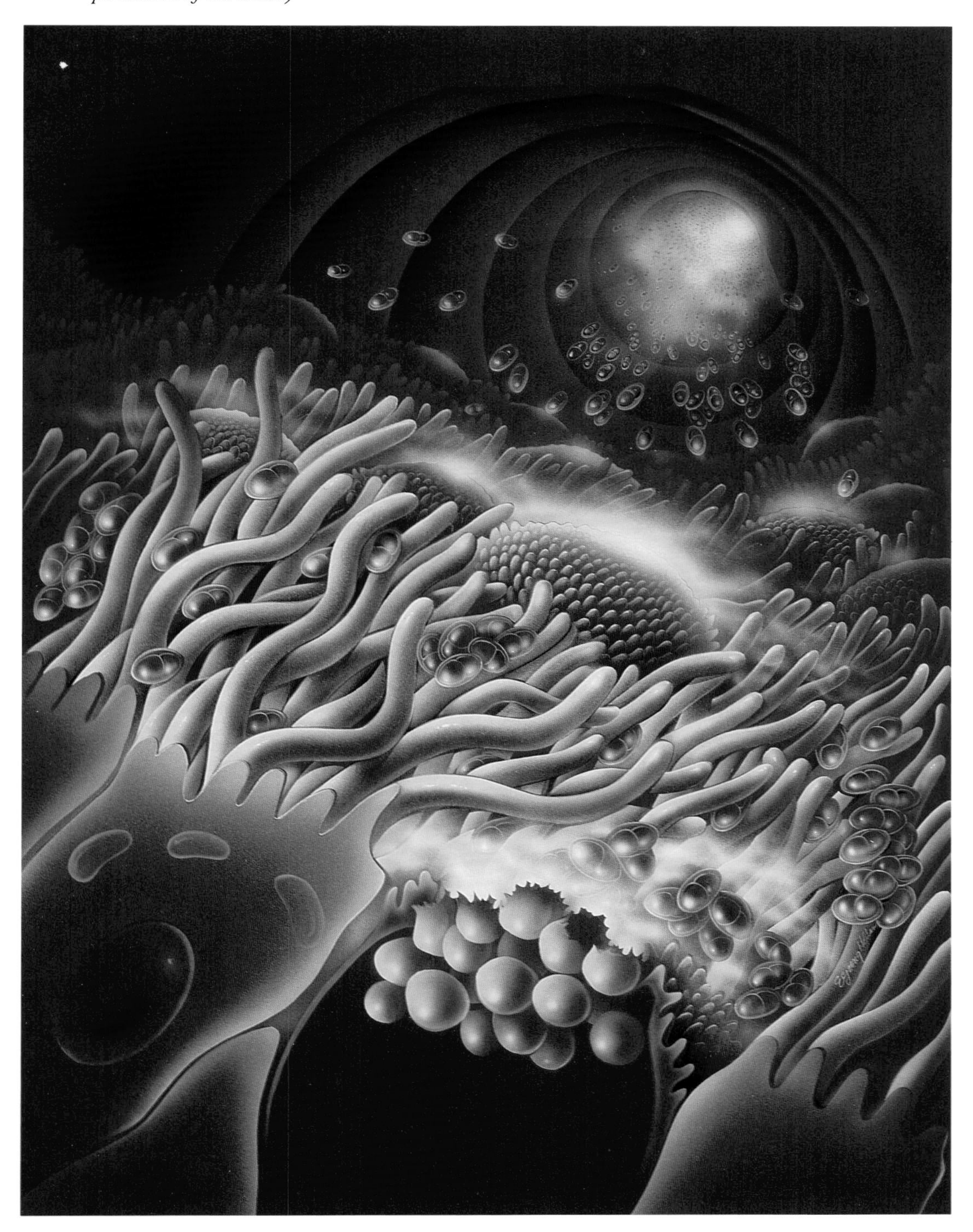

ache) paints are applied too thinly or onto a flexible surface, they may crack or peel.

Airbrush

The airbrush can make a job or ruin it. Considerable practice is necessary to master the technique, but if soft, smoothly blended tones are desired the airbrush is undeniably the tool to use. Any finely ground water-soluble pigment or aniline dye can be used in the airbrush. These can be either transparent or opaque. Hot-press illustration board is an ideal surface for airbrush renderings of mechanical subjects, such as instruments, although a variety of effects can be achieved on many different types of grounds. It is important to test each spray on a swatch of the ground being used before it is applied to the artwork.

Many illustrators avoid using the airbrush because of the time required to cut masks or friskets around areas to be painted, to reload the tiny pigment cups, mix the medium to a proper consistency that will flow through the nib opening, unclog the barrel, and deal with exasperating mechanical adjustments. However, the results can be spectacular, and any errors are easily covered and corrected.

Reviewing Color Proofs

Color proofs must be reviewed by the artist and author before final printing.

Although there may be as many as 25 different kinds of proofs used in composing rooms, engraving departments, pressrooms, and for customer approval, few are useful or meaningful to an author or artist in comparing quality of reproduction to original artwork. Most proofs are for shop use for the engraver's and printer's personnel. The author may be asked to approve customer's proofs, which are printed on enamel stock for the purpose of showing the completed makeup. Other proofs that might show trouble areas are progressive proofs. These will show (1) impression of each color, (2) sequence of colors, and (3) the effect of combining one color with another. Press proofs, showing the color subject as it will appear in print, can also be supplied, but are generally more expensive to produce (see Chapter VIII).

The illustrator of the artwork, who is best qualified to judge the fidelity to the original, must be consulted when proofs are submitted for review. It should be noted that the artwork bears the illustrator's name and not that of the author or the printer. The author is not as likely to be the best judge of problem areas on a color proof, nor to best suggest how to correct them. The artist who knows the language of the printing trade is most qualified to give directions for correcting less than high-quality reproduction.

Computer Graphics

The graphics computer is fast becoming the artist's most exciting creative tool. Electronic technology has advanced to a stage that enables graphics in full color and black and white to be created in significantly less time than required for conventional production at the drawing board. The creative possibilities offered by a dedicated, high speed graphic arts computer, with high resolution and custom designed software, are limited only by the buyer's budget and/or imagination (see Chapter V).

Today's more sophisticated computer graphics systems permit the artist to draw freely on the computer's art terminal, creating images, manipulating their size and position, and developing text, designs, and special effects from a selection of 16 million distinct colors or black and white.

An integral feature of some computers is the capability to receive raw data and to then accurately and automatically plot charts and graphs of any design. These computers can also automatically place statistically accurate curves and standard error bars. Special effects that enhance dimension, create perspective, and sequential spectrum colors can give the effect of a painting or lithograph.

Once artwork is created at the art terminal it is sent to the computer's film recorder, which exposes the images to either color or black-and-white film. The film is then processed in a photo lab to yield color slides or black-and-white prints. In black-and-white prints, colors are converted to screens or textures and the images are camera-ready for publication as line art. If a laser printer is hooked up to the computer, black-and-white laser prints can also be produced. These are useful for author's proofs and for filing, if not also for photographing.

All visual materials created on a graphics computer can be saved on floppy or hard disks and filed indefinitely for quick retrieval and modification.

When it is desirable to publish a computer-generated graphic in color, the slide can be provided for color separation and the printer proceeds with normal color print production.

Ownership

If the artist is employed solely by the author or an institution to provide illustrations for any and all of their uses, and is paid a regular salary with appropriate deductions made by the employer for tax withholding, etc., then, as an employee, the artist has no vested right in the artwork he/she produces.

FIG. III–47: *A. Computer generated illustration of anatomy of the shoulder girdle, created for slide and color print. B. Computer generated bar graph used for slide. C. Conversion of B. to black and white for use as camera ready copy for publication. (All images created on the SlideTek computer graphics system.)*

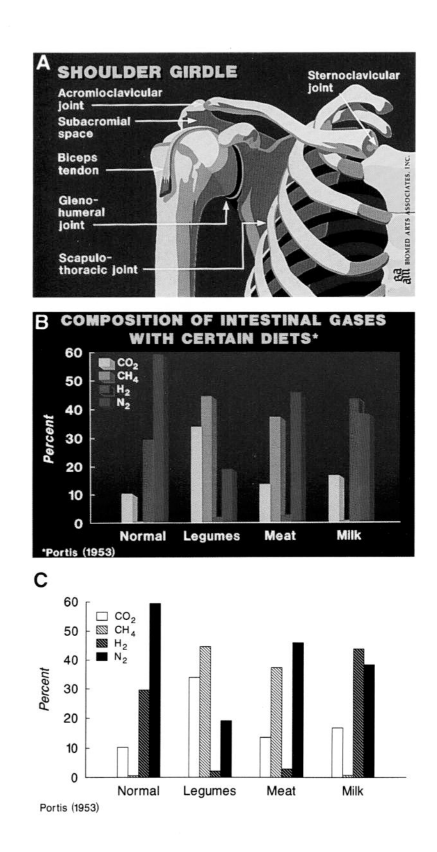

On the other hand, an independent artist who produces specific illustrations for an author is, by law, the owner of the artwork in the absence of a written agreement to the contrary. Before beginning work on any illustration all parties involved (artist, author, publisher) must reach agreement on payment (one-time payment, progression payments, royalties, purchase or no-purchase of ownership, re-use fees, foreign rights, etc.).

The artist should sign the artwork with © preceding his/her name if copyright has not been assigned to author or publisher. The artist may also wish to stamp the back of the illustration with a notice stating: "For one-time reproduction rights only. Artwork remains the property of the artist (or institution). Promptly return unaltered to (name and address of artist or institution)." (see Chapter XI, "Legal and Ethical Considerations".)

Agreements and Contracts

Ideally, any agreement between the artist, author, and/or publisher should be in writing. It may be written as a letter of intent which the parties sign, or as a formal contract. Sample contract forms are available from the Association of Medical Illustrators and the Graphic Artists Guild. Regardless of how the parties wish to express their confirmation of agreement, the following points should be defined:

1. Date of agreement and the names of the artist and client.
2. A clear description of the assignment and specifications, if any.
3. A schedule of delivery stating when preliminary work and finished work are to be delivered to the client, and how (by hand, courier, mail).
4. The artist's fee and the terms of payment.
5. An estimate of possible additional expenses, such as travel, lodging, long distance telephone calls, unusual materials, etc.
6. Cancellation fees are generally calculated as percentages of the total fee for the entire project and are paid to the artist in the event of cancellation or breach by the client. For example, 10% to 25% of the total fee would be paid for cancellation of the project before sketches are completed, 50% after sketches are done, and 100% after the final work is completed.

Once specific terms are agreed to by artist and client, other terms should be addressed. These include:

1. The granting of rights by the artist to the client for reproduction of the work.
2. Limitation of these rights, if any, such as first use only in the first edition in North America.
3. Identification of other limited uses, such as promotion and the medium in which the work may be advertised.
4. Ownership, vested in the artist on completion of the work, may be retained or sold for an additional fee.
5. If the artist retains ownership, an agreement should be made that the artist will not sell the work to a competitor of the client for a specified period of time.

6. The client should agree to give name credit to the artist in print and place a copyright notice adjacent to the artwork in the form of "© artist's name, 19__."
7. The client should agree to indemnify the artist against all claims, expenses, and attorney's fees arising from client's use or sale of the reproductions.
8. Artist should warrant that the artwork is original and does not infringe on any other copyright.
9. It should be stated that disputes arising out of the agreement will be submitted for binding arbitration and that the non-prevailing party shall pay the incurred costs, fees, and interest.

CHAPTER IV
GRAPHS AND MAPS

Displaying Data

Data are facts that have been collected systematically. In science, data are often numerical and thus are well suited to being displayed in tables, graphs, or maps. These forms are more concise than the written word. They can also be more forceful—if the form suits the purpose and if the display is thoughtfully designed and skillfully executed. This chapter presents guidelines for designing some commonly used displays of data.

A few general principles hold for all these forms of display. However, because each form organizes data in a different way, the bulk of this chapter is devoted to guidelines for designing the individual types of each form. Guidelines for two other commonly used visual displays, algorithms and flow charts, are also included.

The guidelines are intended to be applied flexibly and to encompass a range of possibilities. Thus, in the illustrations in this chapter details vary from one graph or map to another. Not all possible variations are represented.

Definitions

Table: A table is a systematic arrangement of data, usually in rows and columns.

Graph: A graph is a diagram that shows amounts, frequencies, trends, or relationships of data. The term "graph" is sometimes restricted to line graphs, which show relationships between two or more variables. Graphs that show amounts or frequencies of one variable are then referred to as charts. In this chapter, the broader definition of graph is used.

Distribution map: A distribution map is a map that displays spatial relationships of data.

Algorithm: An algorithm is either a diagram of a procedure that leads, by a series of choices, to a correct answer or a diagram of a method of breaking a complicated decision-making sequence into its components.

Flow chart: A flow chart is a diagram that displays processes, sequences, or systems.

Selecting the Form for a Display of Data

Before selecting the form for display of data, the author must decide on the purpose of the display. The purpose of the display dictates the form. The author should select:

- a table when exact values are important;
- a graph when trends or relationships are more important than exact values; when the meaning of the data needs to be quickly and forcefully expressed; when hidden relationships need revealing (for example, an S-shaped curve is more apparent from a graph than from a table);
- a distribution map when the location of data is more important than actual numerical values;
- an algorithm when the succession of steps used in problem solving needs to be displayed;
- a flow chart when processes, sequences, or systems need to be presented in an organized fashion.

General Principles for Displaying Data

Accuracy

The value of any display of data depends on the integrity and care with which the data were collected and analyzed. No table, graph, or map, however carefully designed and beautifully executed, can redeem poor or inaccurate data.

Selectivity

The number of displays should be limited to the fewest that will cogently express the message of the work. It is more difficult for the reader to pull together the message from 17 tables, graphs, and maps than from five. Similarly, there is a limit to the amount of information that can be conveyed intelligently by any one table, graph, or map. Only the information that is necessary for conveying the message should be included. For example, data from one experiment can be displayed, but data from a parallel experiment yielding similar results can be summarized in the text.

Nonredundancy

Data should be presented only once. The same data should not be presented in both a table and a graph, or in both a table and a map.

Consistency

Similar data should be displayed in similar form. Alternating between graphs, maps, and tables merely for variety is distracting.

Focus

A table, graph, or map should make a point. The point should be apparent from the design of the display and, whenever possible, should be stated in the title. One point is usually enough. However, two or three points may be made if the display remains simple.

Clarity and Simplicity

A table, graph, or map should be clear and simple. Undue complexity of data or of explanatory detail, awkward word choice, or cryptic abbreviations make the point difficult to grasp. Similarly, in graphs and maps, careless choice and arrangement of design elements such as labels and keys cause visual distraction. If the science is complex, the message should be conveyed as simply and clearly as possible. If the science is simple, it should not be made to appear complex.

Visual effectiveness

Graphs and maps are visual media. For graphs and maps to be visually effective, words should be kept to a minimum. Information that cannot be conveyed visually should be omitted.

Convincingness

A graph or map should be visually convincing. For example, if the point of a graph is that a variable increased, the increase should be readily apparent.

Independence

A table, along with its title and footnotes, and a graph or map, together with its legend, should be understandable without reference to the text.

Tables

The table is a form of data presentation that is visually organized yet still exact, a sort of halfway point between text and graph. It allows the writer to present even large amounts of information—numbers and words—in a form that is readily accessible to the reader. A table is not considered an illustration.

Guidelines to the conventions for constructing tables can be found in the Council of Biology Editors Style Manual, Fifth Edition.

Graphs

Graphs are used to show amounts, frequencies, trends, or relationships of data. The type of graph used depends on both the purpose of the graph and the type of data.

Common Types and Purposes

To show a relationship between two ratio-scale variables, such as a trend with time or the dependence of a response on the concentration of a stimulant, a line graph is used. (Ratio-scale variables have constant intervals between successive units and a zero point that has a physical meaning, for example, weight, volume, pressure, time, concentration.)

To show whether there is a correlation between two ratio-scale variables, a scatter diagram ("scattergram") is used.

To show amounts or frequencies for nominal-scale or ordinal-scale data, a bar graph (also called a bar chart) is used. (Nominal-scale data have no intervals. Nominal-scale data are subtypes of a class and are distinguished qualitatively, for example, types of cells, types of drugs. Ordinal-scale data have no constant intervals but they have relative ranks, for example, responses graded from least to most.) Bar graphs are also used for ratio-scale variables, particularly when data were collected at unequal intervals.

To show a frequency distribution of data for a ratio-scale variable, either a histogram or a frequency polygon is used.

To show the components of a whole, either a component bar graph or a pie graph (also called a pie chart) is used.

Guidelines for Preparing Graphs

Line graphs and scattergrams

DEFINITIONS

A line graph is a graph that has two axes, at right angles to each other, which represent the scales of two ratio-scale variables (for example, weight, volume, pressure). The axes form a coordinate grid, on which a relationship between the two variables is shown. One variable may be dependent on the other. An independent variable is a variable for which the investigator sets a value or a range of values; a dependent variable is a variable that changes as the independent variable changes. For example concentration (dependent variable) may change with time (independent variable).

Conventionally, the independent variable is graphed horizontally, along the X (abscissal) axis, and the dependent variable is graphed vertically, along the Y (ordinal) axis (Fig. IV–1). Thus, the graph shows what happens to Y as X changes. The relationship in a line graph is depicted by curves, by data points, or by both.

Three common types of line graph are the linear (arithmetic) scale graph, the semilogarithmic scale graph ("semilog graph"), and the logarithmic scale graph ("log-log graph"). In linear scale graphs, equal distances represent equal quantities everywhere on each axis (Fig. IV–1). In

FIG. IV–1: *A linear-scale line graph. Both the Y (vertical) and the X (horizontal) axes are in linear scale. The Y axis shows the dependent variable (trehalase specific activity) and the X axis shows the independent variable (time). Note also the following details: Tick marks are at equal intervals and are all the same length. Tick marks are numbered at conventional intervals. Not every tick mark is numbered. Tick marks are visible but not obtrusive. Tick marks point out, so that the tick mark at 200 will not run into the curves. Two zeroes are used to label the point where the axes meet. The numbers along the Y axis are oriented horizontally, the same as those along the X axis. The Y-axis label is parallel to the Y axis and reads from bottom to top. In the Y-axis label, the unit of measurement (in parentheses) is below the name of the variable. In addition, square data points are kept away from circles as much as possible. Symbols for data points are about three times as wide as curves. Curves are broken around black symbols. The outline of white symbols is prominent. The curves were fitted by eye, so they are drawn as straight lines between each pair of data points. One curve is interrupted where it crosses another (except where the angles are especially shallow). Only the symbols are different, not the line patterns. (Reproduced, with permission, from Thevelein JM, den Hollander JA, Shulman RG. Proc Natl Acad Sci USA 79:3503, 1982.)*

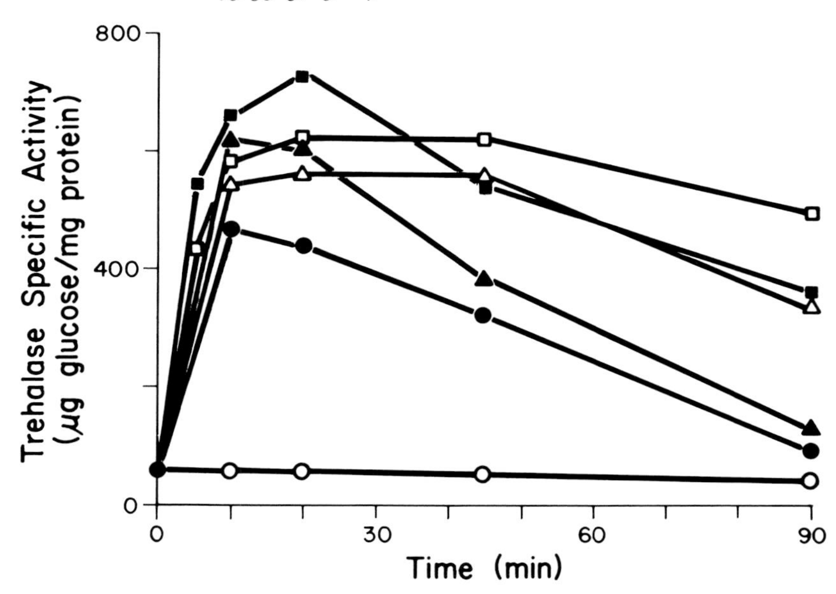

semilog graphs, one axis is in linear units and the other is in logarithmic units (Fig. IV–2). In log-log graphs, both axes are in logarithmic units (Fig. IV–3).

A scattergram is a graph drawn on a coordinate grid on which each

FIG. IV–2: *A semilog graph. The Y axis is in logarithmic scale and the X axis is in linear scale. Note also the following details: The tick mark at the beginning of each logarithmic cycle is longer than the others. Because more than two complete cycles of logarithms are shown, only the tick marks at the beginning of cycles are numbered. Logarithmic scales do not contain 0, so the axes do not meet at (0, 0). Note also that scale numbers along the Y axis are aligned on the decimal point. A zero is placed before scale numbers that are less than 1. The Y-axis label is centered perpendicularly above the Y axis. The curves were fitted by an equation, so the curves are smooth and do not pass through every data point. (Reproduced, with permission, from Yoakum GH, Yeung AT, Mattes WB, Grossman L. Proc Natl Acad Sci USA 79:1766, 1982.)*

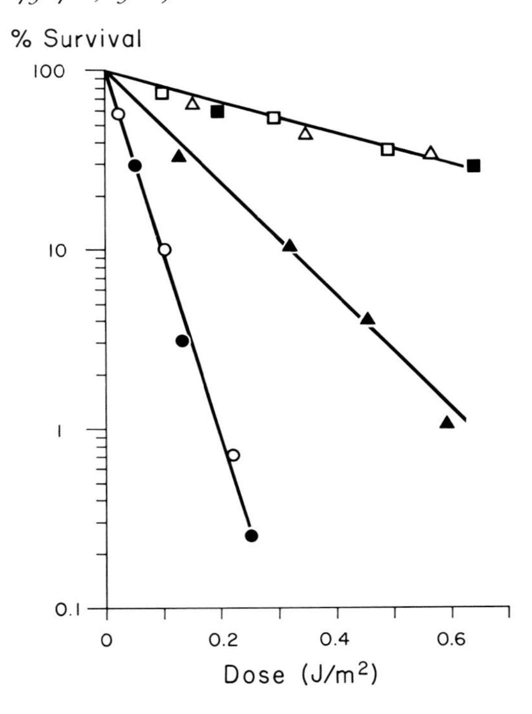

individual event or value is represented by a data point (Fig. IV–4). Whether the variables shown on the axes are correlated, and, if so, to what degree, can be determined by trying to fit a mathematical function to the data points. For example, a linear function can be shown by fitting a straight regression line to the data points (Fig. IV–4). One way that the strength of a linear association can be estimated is by determining the correlation coefficient (r) (see Fig. IV–4).

FIG. IV–3: *A log-log line graph. Both axes are in logarithmic scale. Note also: The X-axis label is centered under the X axis. The Y-axis label is centered parallel to the Y axis. The axes are shifted apart slightly to make the data point that falls on the Y axis visible. Only one symbol is needed for the data points, so a dark circle is used. (Reproduced, with permission, from Helander HF, Durbin RP. Am J Physiol 243:G297, 1982.)*

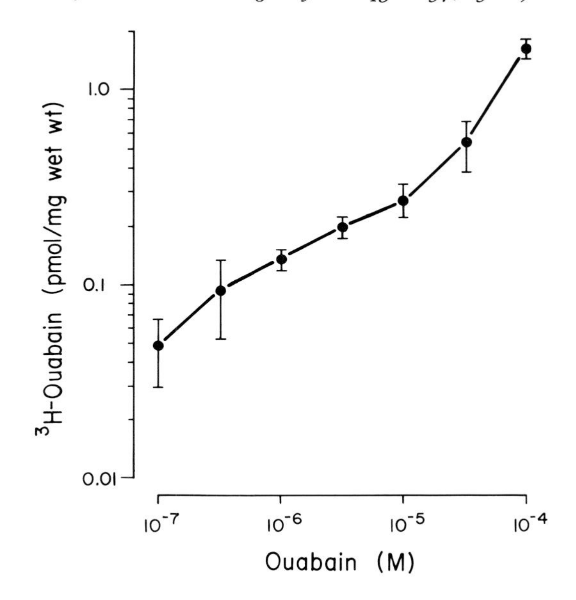

GUIDELINES FOR LINE GRAPHS AND SCATTERGRAMS

The axes

An axis is a reference line in a coordinate grid. Every axis must be scaled and identified by a label.

Placement of the axes

In a graph showing any single quadrant, the axes are conventionally at the left and the bottom (Figs. IV–1–4). In a graph showing more than one quadrant, the axes should meet at the origin so that it is immediately apparent whether the data are positive or negative (Fig. IV–5).

Scales and scale numbers

The scales on each axis, indicated by tick marks, must be accurate. That is, every interval on a linear scale and every cycle on a logarithmic scale must be superimposable on every other interval or cycle along that scale.

The purpose of tick marks is to identify the type of scale, not to identify each data point. Therefore, on linear scales, tick marks must be placed at equal intervals (as in Fig. IV–1). Unequal intervals, such as 0, 4, 30, 37°C or 0, 2, 5, 10, 30 minutes, should not be used.

FIG. IV–4: *A scattergram. The intervals on both axes are the same. The excellent fit of the straight regression line (r = 0.976) indicates that the two variables have a strong association. Note also: Data points that overlap are drawn overlapped. One zero is used to label the point where the axes meet. A key is used to identify the two populations of data. The key is placed in an area within the rectangle of the graph where there are no data; the key is not boxed. Only one symbol is used in the key to identify each data set. Line weights are as follows: boldest, regression line; less bold, axis labels; least bold, letters in the key and in the statistical note, axes, tick marks, and scale numbers. (Reproduced, with permission, from Brill DM, Wasserstrom JA. Circ Res 58:109, 1986.)*

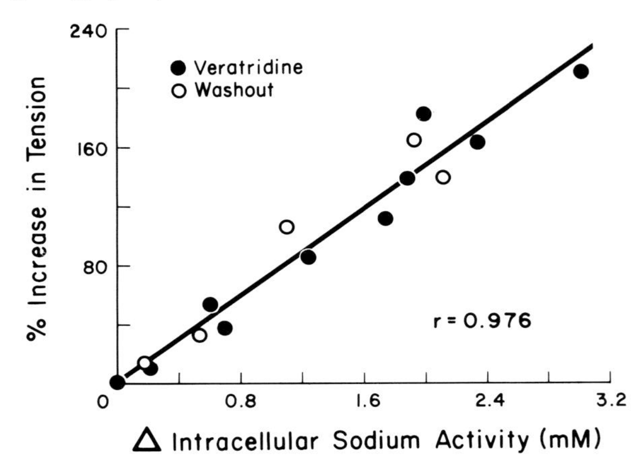

The tick marks must be labeled, usually with numbers, but sometimes with letters standing for the names of months. The first number should be at the beginning of the axis; other numbers should be at equal intervals (Fig. IV–1), not just at the end of the axis. When tick marks on a linear scale are closely spaced, only every second, third, fourth, or fifth tick mark needs to be numbered (see Figs. IV–1, 4). Numbering should be at conventional intervals, for example, 0, 2, 4, 6 . . . ; 0, 30, 60, 90 . . . ; 0, 5, 10, 15, 20 . . . ; 0.0, 2.5, 5.0, 7.5, 10.0 . . . ; but not 1, 4, 7, 10, 13 . . . or 6.0, 6.6, 7.2. . . . It is not necessary to have a scale number at the end of the axis (see Fig. IV–2, X axis). However, the axis should end at a tick mark, to permit easy estimation of the value of data points. Numbers along the Y axis should be aligned on the decimal point (whether or not a decimal point is actually present) (see Fig. IV–2, Y axis). In a graph showing more than one quadrant, if scale numbers placed along an axis would not be visible, they can be placed at the left (for the Y axis) or along the bottom (for the X axis) (see Fig. IV–5). In this case, the tick marks should be repeated along with the numbers.

Linear scales should have enough tick marks to permit readers to esti-

FIG. IV–5: *A line graph showing two quadrants. The X axis is not drawn along the bottom. Rather, the axes meet at the origin. However, the scale numbers would not be visible along the X axis, so they are placed along the bottom, as are duplicate tick marks. Note that the three curves cross and therefore are distinguished by different line patterns. The line patterns are of equal weight, so the three curves look equally important. The curves are identified by numbers. Note also the multiplier in the Y-axis label; there is no convenient way to avoid a multiplier in this case. Also, the Y-axis label is placed perpendicular to the Y axis, close to the scale numbers, but not so close that it protrudes into the imaginary rectangle occupied by the scale numbers. (Reproduced, with permission, from Nakano M, Iwamaru H, Tobita T, Yang JT. Biopolymers 21:805, 1982.)*

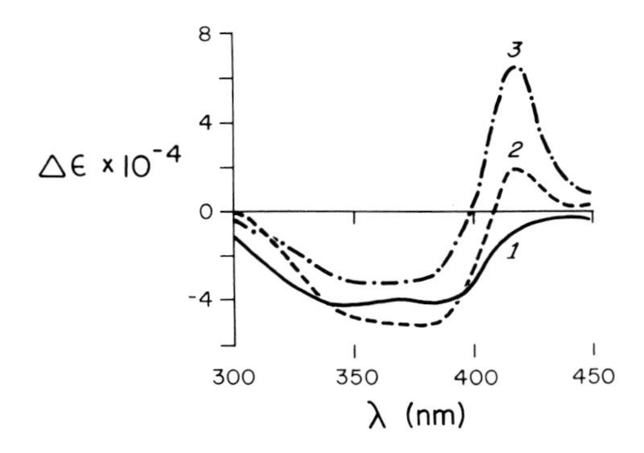

mate the value of data points easily, but not so many that the axis is cluttered. All tick marks on a linear scale should be the same length.

There are no fixed rules for the size of the intervals along an axis. When deciding what size interval to use, watch the impression the resulting curve makes. Too steep or too shallow a curve can create a misleading impression of what the data mean. One general rule, however, is that in a scattergram that compares two ways of measuring the same variable, the intervals on both axes should be the same size. If possible, the intervals on both axes should also be the same size in other scattergrams (see Fig. IV–4).

Log scales are easiest to identify if all the intervals are indicated by tick marks, and the tick mark at the beginning of each cycle is longer than the others (Fig. IV–2, Y axis). When two or more complete cycles are shown, only the tick marks at the beginning of each cycle need to be numbered (see Fig. IV–2). When fewer than two full cycles are shown, numbering is usually at 1, 2, 5; 10, 20, 50; and so on.

The point where the axes meet must be numbered. When the axes meet at (0, 0), both zeroes can be labeled (Fig. IV–1), or one zero can be used to label both (Fig. IV–4). When the axes do not meet at (0, 0), two numbers should be used to identify the meeting point (Fig. IV–6).

Logarithmic scales have no 0, so the axes in a semilog graph or in a log-log graph cannot meet at (0, 0) (Figs. IV–2, 3).

A zero should be placed before scale numbers less than 1 (see Fig. IV–2, X axis) because the decimal point is easy to miss if there is no zero.

Numbers along the Y axis are oriented horizontally, just like those along the X axis (see Figs. IV–1–5).

Tick marks may point in or out, but they should not cross the axis. Tick marks that point in direct the eye toward the data. Tick marks that point out keep the face of the graph clear and are particularly useful when data fall on the axis (see Fig. IV–1). Tick marks should be long enough to be visible but not obtrusive (Figs. IV–1–5).

Using multipliers of the scale numbers as a means of avoiding excess zeroes is confusing and should be avoided where possible. The best way to avoid excess zeroes in scale numbers is to change the unit of measurement (for example, 0.001 ml becomes 1 μl).

Axis labels

Each axis must be clearly labeled with both the name of the variable and the unit of measurement. Conventionally, the name of the variable is given first and the unit of measurement (in International System [SI] abbreviations) is given immediately after or below it in parentheses, for example, "Temperature (°C)." Exceptions to the unit following the name of the variable are variables having dimensionless quantities, such as counts (for example, number of eggs), ratios, and pH.

Axis labels should be brief. If necessary, a brief label can be explained more fully in the legend. For example, for the Y-axis label "[^{3}H] 2-DG Uptake (pmol/10 cells)," the title in the figure legend identifies 2-DG as 2-deoxyglucose and the cells as thymocytes ("Fig. 1. Calcium dependence on the effect of T_4 on the uptake of [^{3}H] 2-deoxyglucose by rat thymocytes"). Standard abbreviations can be used to keep axis labels short (for example, % for "percentage" or "percent").

Axis labels should be placed outside the graph. The X-axis label is customarily centered under the X axis (see Fig. IV–2). The Y-axis label can be centered perpendicularly above the Y axis (as in Fig. IV–2), placed perpendicular and to the left of it (slightly above the middle of the Y axis), if the label is short (as in Fig. IV–5), or centered parallel to it (as in Figs. IV–1, 3). The parallel label is conventional, and is best when horizontal space is limited. However, perpendicular labels are easier to read. Whether perpendicular labels should be placed above or to the left of the Y axis depends on the amount of horizontal space available. In both parallel and perpendicular labels, the unit of measurement can be placed below the name of the variable (see Fig. IV–1, Y-axis label). Parallel labels should read from bottom to top (Fig. IV–1).

Axis labels should be placed close to the scale numbers, but not so close that they protrude into the imaginary rectangle occupied by the scale numbers (see Fig. IV–5, Y-axis label).

Shifting the axes
Normally, the axes should meet. However, to make data that fall on an axis more easily visible, the axes may be shifted apart slightly (as in Fig. IV–3).

Starting the scales near the data
To save space, scales should be started close to the first data point, as in Fig. IV–6, not at (0, 0). In linear scale graphs, to make clear that the axes do not meet at (0, 0), the axes may be shifted apart slightly (as in Fig. IV–6). This technique is not necessary in semi-log graphs (see Fig. IV–2) or in log-log graphs, whose axes cannot meet at (0, 0).

Axes should not extend far beyond the data. Excess white space makes the data seem unimportant.

Breaking the curves
As a rule, curves should not be broken. The purpose of presenting data on a graph is to show a pattern, and breaks interrupt or distort the pat-

FIG. IV–6: *A line graph in which the scales begin close to the first data point. Note that the axes are shifted apart slightly to indicate that they do not meet at (0, 0). Note also: A horizontal bar is placed above the data to indicate the duration of an event (stimulation of the carotid sinus nerve). The curves are identified by brief labels, each clearly referring to the appropriate curve. The labels are placed horizontally, are not joined to the curves by leader lines or arrows, and are not boxed. Since the curves are labeled and do not cross, different symbols are not needed to distinguish the curves. Error bars are used to indicate standard deviations. (Reproduced, with permission, from Lawson EE, Long WA. J Appl Physiol 56:1614, 1984.)*

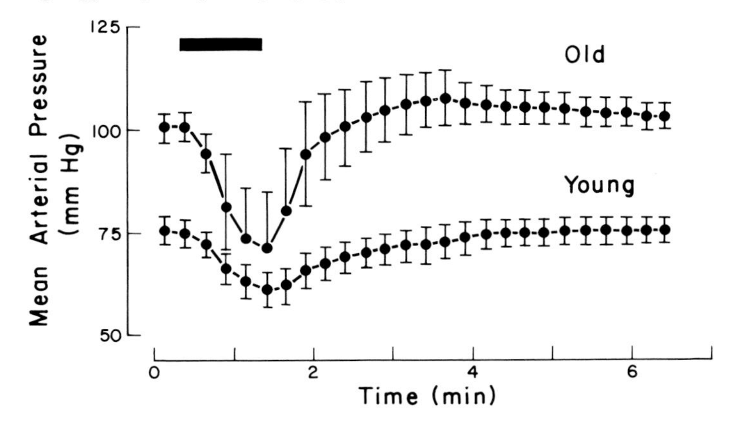

tern. However, in a graph that covers a wide range of data, you might choose to break a curve to eliminate portions of a curve whose slope is not changing. Such breaks do not create problems for the reader if the break is clearly indicated. To indicate a break, draw two short, dark, parallel, slanted lines, one at each side of the break. Draw similar short, dark, parallel, slanted lines at the corresponding part of the axis. Thus the reader will be aware of the break at first glance and will not be deceived into thinking that an uninterrupted curve is shown.

FIG. IV–7: *Two ways of drawing a line graph that covers a wide range of data. A. A line graph in which the curves are broken, thus distorting the trends. To partly alleviate the distorted impression, the breaks in the curves, and in the corresponding part of the axis, are indicated by pairs of short, dark, parallel, slanted lines, and the curves are drawn at appropriate levels before and after the breaks. Also, the scale is the same before and after the break. B. The same data drawn on a semilog graph, which shows the trend more truthfully. (Reproduced, with permission, from Lahiri S, Mokashi A, Mulligan E, Nishino T. J Appl Physiol 51:55, 1981.)*

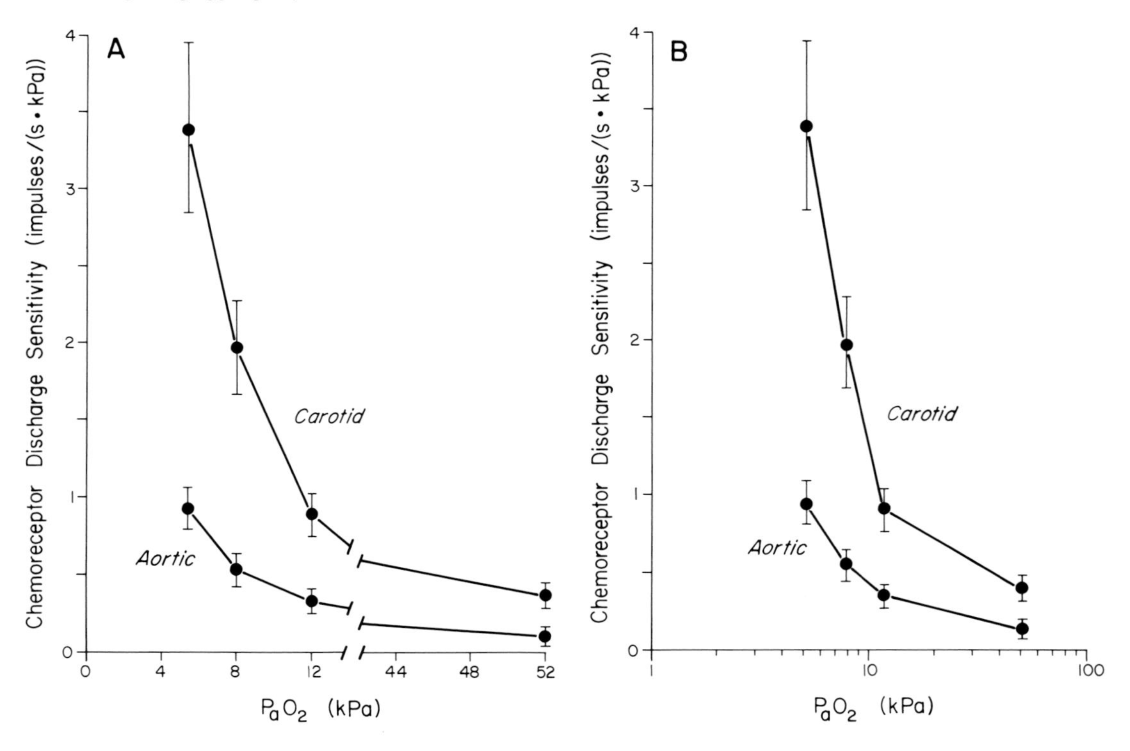

If a break occurs where the slope of the curve is changing, the curve will be distorted. This distortion falsifies the curve. To partly alleviate the false impression, draw the curve at appropriate levels before and after the break (Fig. IV–7A). Also, draw short, dark, parallel, slanted lines to indicate both the break in the curve and the corresponding break in the axis. To ensure that the slopes before and after the break are accurate in the final graph, first draw the full curve, and then cut out the portion you want to omit and splice the remaining portions. A better solution than breaking the curve is to change to a semilog graph (Fig. IV–7B, inset).

The scales before and after a break must be the same (see Fig. IV–7A). If they are not, the overall curve is not consistent. (The same is true even when there is not a break.) Although the reader might eventually discover the different scales, it is best to present an accurate visual image on paper so that the reader is not misled.

More than one Y axis

A graph that has more than one Y axis is difficult to read. When more than one Y axis is needed, all Y axes should extrapolate to the same zero point, and the labels for all the Y axes should read in the same direction (from bottom to top) (Fig. IV–8). If the labels are brief, they can be centered above the Y axes. It is helpful to include the data point or line pattern in the Y-axis labels to indicate which data relate to each Y axis (Fig. IV–8). One data point in each Y-axis label is sufficient. If a graph having more than one Y axis becomes too complex, each graph can be presented separately in a composite figure (see "Sets and Composites" below).

Framed graphs

It is generally not necessary to frame the graph, i.e., to put axes on all four sides. If axes are needed on all sides to help the reader estimate the value of data points, all axes must have scales. The intervals on the scales of corresponding axes (for example, on both of the X axes) must be identical.

Symbols for data points

When data points for individual or mean values are represented, polygon symbols should be used. The recommended symbols are ●, ○, ▲, △, ■, □. In addition, ▼, ▽, ◆, ◇ are available, but they are not distinctive and therefore should be avoided. If they are used, they should be carefully oriented so that they can be distinguished from ▲, △, ■, □. X, +, and * are available, but they should not be used because they are not as prominent as polygon symbols; also, they can disappear into the line pattern. ⊙ and similar symbols should not be used because they do not reproduce clearly.

FIG. IV-8: *A graph having two Y axes. Both Y axes extrapolate to the same zero point. Labels for both read from bottom to top. The appropriate data point is included in each Y-axis label. (Reproduced, with permission, from Helander HF, Durbin RP. Am J Physiol 243:G297, 1982.)*

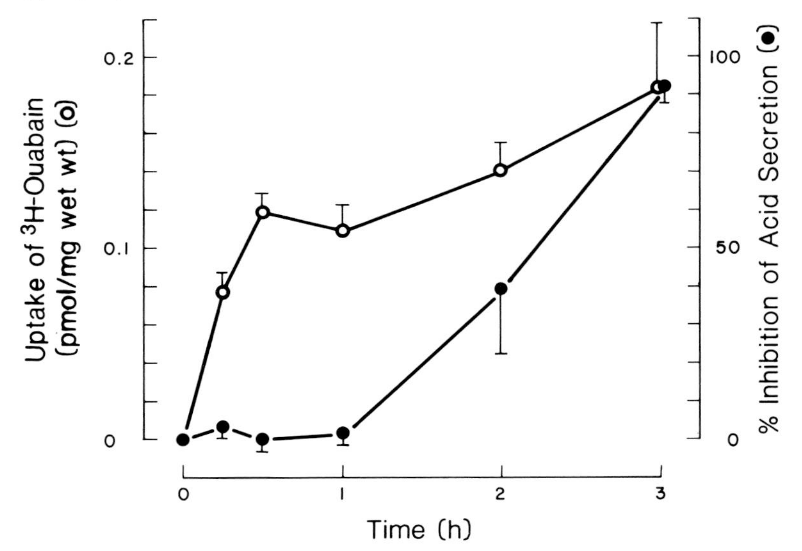

For data points that overlap, the symbols should be drawn overlapped (see Figs. IV-1, 4). Data points that coincide are difficult to represent. One solution is to use a single symbol to represent all the data points that coincide, as in Fig. IV-1, in which a black circle at time zero represents six data points.

Symbols are distinguished by color (black or white) and by shape. The easiest symbols to distinguish are black from white. The easiest shapes to distinguish are circles from triangles. Therefore, if two symbols are needed, ● and ○ are first choice. If three symbols are needed, reasonable choices are ●, ▲, ■; ●, ○, ▲; or ●, ○, △, depending on what the symbols represent (see next paragraph) and whether the ● and the ■ fall next to each other. Circles should not fall next to squares because these two symbols are difficult to distinguish, especially after the graph is reduced to publication size. If four symbols are needed ●, ○, ▲, △ are good choices, alternated by color if possible. If more than four symbols are needed, squares should be used, and care should be taken to keep squares away from circles (for example, as in Figs. IV-1, 2).

Color can be used symbolically: for example, ○ no treatment, ● treatment. Color and shape can both be used to group data: for example, ○ before and ● after one treatment, and △ before and ▲ after another treatment.

Black symbols are more prominent than white symbols (see Fig. IV–1). Circles are the most like points. Therefore, if only one symbol is needed, black circles should be used (as in Figs. IV–3, 6, 7).

All symbols should look the same size.

To ensure that data points will be visible after the graph is reduced to publication size, the symbols should be about two to three times the width of the curve (see Figs. IV–1–3). The curve should be broken around black symbols (as in Figs. IV–1–3) so that after reduction the curve does not blur the shape of the symbol. The outline of white symbols should be bold enough that the white symbols look nearly as prominent as the black symbols (Figs. IV–1, 2).

Symbols represent a point, so symbols should not be drawn too large. The more data points, the smaller the symbols need to be. However, symbols should not be so small that they look like part of a line pattern.

Curves

Curves can be fitted to data points by eye or by an equation. Curves fitted by eye should be drawn as straight lines between each pair of data points (Fig. IV–1). Curves fitted by an equation should be smooth and probably will not pass through every data point (see Figs. IV–2 and IV–9). Curves should not be extended before the first data point or after the last data point. However, the author may occasionally want to extrapolate. In this case, a different line pattern (usually a dashed or dotted line) should be used for the extension, and this pattern should be identified as extrapolation in the figure legend.

Curves should be the heaviest lines on a graph.

If curves need to be distinguished, they should be distinguished by lines of different patterns but equal weight (Fig. IV–5), not by lines of different weights, because heavier lines seem more important.

When different symbols are used to distinguish the data, different line patterns are not usually necessary, even if the curves cross (Fig. IV–1). Too many line patterns can be confusing.

The number of curves that can be presented clearly on one graph depends on the location of the curves and on the purpose of the graph. Generally, five or six curves are enough for one graph. However, if five curves cross frequently and the reader is expected to follow each one, five curves could be too many. Conversely, if all curves are well spaced or if the purpose of the graph is, for example, to show several similarities and one outstanding difference, then more curves can be included without jeopardizing clarity.

When curves cross, lines can be interrupted so that one curve clearly passes in front of the other, as in Figs. IV–1, 5. Thus, each curve is easy to follow. However, if the angles of the crossed curves are shallow, interrupting the curves is unnecessary (see Fig. IV–1, the second and third curves from the top, at 60 min).

Curve labels and keys

The clearest way to identify curves is by a brief label (Figs. IV–5–7). If space does not permit using curve labels, second choice is to use symbols or line patterns to distinguish curves and to define the symbols or line patterns in a key placed on the face of the graph (Fig. IV–4). However, if labels are not feasible and a key will clutter the graph, the symbols or line patterns should be identified in the legend (third choice).

When curves that do not cross are identified by labels, different symbols or line patterns are not necessary (Figs. IV–6, 7).

A label identifying a curve should be placed close to the curve, in a spot where the label clearly refers only to the appropriate curve and where the label is easy to see (as in Figs. IV–5–7). Curve labels should be within the rectangle implied by the axes, not at one side of the graph, because extending the width of the figure decreases the size of the published graph. Curve labels should be oriented horizontally (see Figs. IV–5–7), not at an angle. Arrows or leader lines should not be used to join labels to curves because they clutter the graph. Labels should not be boxed because boxing calls too much attention to them.

Keys should be placed within the rectangle implied by the axes, in a spot where there are no data (as in Fig. IV–4). The key can extend slightly beyond the rectangle of the graph (Fig. IV–13) but should not be completely outside the graph, because then the key will draw the eye away from the data. Keys should not be boxed because the box calls too much attention to the key. The order of symbols or line patterns in the key should follow the order of symbols or line patterns in the graph, if there is an obvious order, for example, at the peak or at one end. Only one symbol is needed in the key, not two symbols connected by a line.

Insets

A portion of a graph can be enlarged and presented as an inset. An inset can also be used to present a related graph (as in Fig. IV–9) or to make a comparison. However, if a graph is important enough to be published, it may be better to put it in a separate panel (see "Sets and Composites," below). If an inset is used, it should be within the rectangle implied by the axes, in a spot where there are no data. The data points, letters, and numbers in insets should be smaller than those in the main graph, but must still be legible after the figure is reduced to fit the journal's column or page. The inset must be described in the legend, just as the main graph is.

Arrows, lines, bars, and shading

Arrows can be used to draw attention to important features of a curve or to identify points of special interest (Fig. IV–17). Arrows or vertical dashed lines can be used to indicate an event. Horizontal bars (Fig. IV–6) or lightly shaded rectangular areas can be used to indicate the duration

FIG. IV–9: *A graph containing an inset, which presents a related graph. In the inset, the data points, letters, and numbers are smaller than those in the main graph, and the curve is less bold. Line weights in both graphs are as follows: boldest, curves; less bold, axis labels; least bold, axes, tick marks, and scale numbers. (Reproduced, with permission, from Murlas C, Nadel JA, Roberts JM. J Appl Physiol 52:1084, 1982.)*

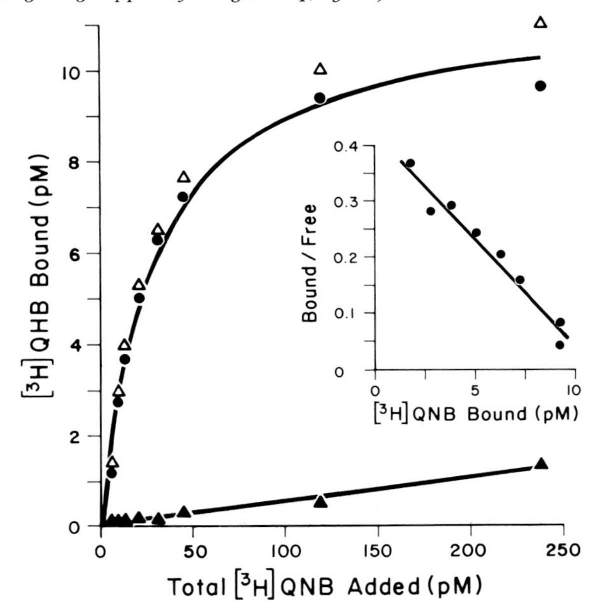

of a treatment or maneuver. So that they will be readily visible, arrows and bars should be placed above the data, not below the data or below the X axis. Arrows and bars should be neither too small or faint nor too big or bold.

Bar graphs

DEFINITIONS

A bar graph (bar chart) is a one-axis graph for comparing amounts or frequencies for classes of a nominal-scale variable, such as different types of cells, or for classes of an ordinal-scale variable, such as graded responses to a stimulus. In addition, a bar graph is sometimes used to compare amounts or frequencies for classes of a ratio-scale variable, particularly when the data were collected at unequal intervals. In a bar graph, the data are compared by means of parallel rectangles ("bars") of uniform widths, each bar representing one class of the nominal- or ordinal-scale variable (for example, one type of cell, one level of response). The length of each bar is proportional to the amount or frequency for its class (Fig. IV–10).

FIG. IV–10: *A bar graph (bar chart) in which the bars are arranged vertically. The axis is at the far left, it begins at zero, and zero is labeled. The baseline is not drawn but is implied by the ends of the bars. Class labels are placed outside the baseline and are centered below each bar. Bars are of equal width and spaces are of equal width; bars are wider than the spaces between them. Bars are outlined boldly and are shaded dark gray. Control data are at the left. Note also the error bars centered at the top of each data bar. Half error bars are used because the data bars are dark. (Reproduced, with permission, from Kato N, Kido N, Ohta M, Naito S, Hasegawa T, Mori M, Agata N, Nakashima I, Kuno T. Microbiol Immunol 27:1043, 1983.)*

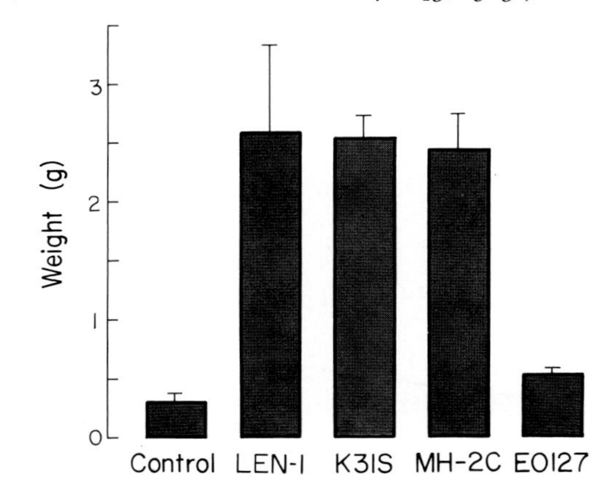

A bar graph can be arranged either vertically or horizontally. The vertical arrangement (Fig. IV–10) (sometimes called a column graph) is the more usual in some scientific disciplines; the horizontal arrangement (Fig. IV–11) is the more usual in other scientific disciplines. The horizontal arrangement is advantageous when long class labels are needed.

Bar graphs lend themselves to presenting comparative data. For example, bars can be grouped (as in Fig. IV–12) or used to show deviations from a baseline (as in Fig. IV–13).

A variation on vertical bar graphs is the one-axis graph that uses data points to show all the individual data (Fig. IV–14) instead of a bar to show the mean. For paired data, the data points are sometimes joined by lines to show the direction of the change, particularly when few data are available (Fig. IV–15). The mean can be shown by a short horizontal line (as in Fig. IV–14) or by a data point (and error bar) placed to one side of the column of individual data points (as in Fig. IV–15).

GUIDELINES FOR BAR GRAPHS

The axis and the baseline

The axis is placed at the far left in a vertical bar graph (Fig. IV–10) and along the top or bottom in a horizontal bar graph (Fig. IV–11).

FIG. IV–11: *A bar graph in which the bars are arranged horizontally. The axis is at the bottom, it begins at zero, and zero is labeled. The baseline is not drawn but is implied by the ends of the bars. Class labels are placed to the left of the axis and, because they are of about equal lengths, they are aligned on the left. (Reproduced, with permission, from Armstrong RB, Delp MD, Goljam EF, Laughlin MH. J Appl Physiol 62:1285, 1987.)*

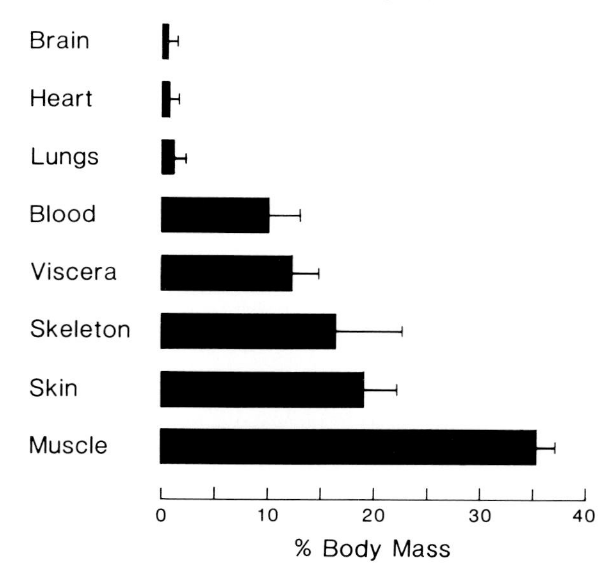

The axis of a bar graph is scaled and labeled the same as the axes of line graphs. The axis of a bar graph must include zero, and zero must be labeled (Figs. IV–10, 11). Because the length of the bar indicates the amount or the frequency, excluding zero from the axis falsifies the comparison shown by the bars. Similarly, breaking the axis and the bars falsifies the comparison, so neither the bars nor the axis should be broken.

The baseline of a bar graph need not be drawn. When the bars are wider than the spaces between them (as recommended below), the baseline is strongly implied by the ends of the bars (Figs. IV–10, 11). However, in bar graphs showing individual values, drawing the baseline may be useful, for example, to identify zero (Fig. IV–14) and thus indicate that some data are negative.

The baseline is not an axis, so if it is drawn, it does not have tick marks (Fig. IV–14).

Class labels

The classes of a bar graph are identified by labels placed outside the baseline. For single bars, class labels are centered below each bar (vertical bar graphs, as in Fig. IV–10) or placed to the left of each bar (horizontal bar graphs, as in Fig. IV–11). In horizontal bar graphs, class

FIG. IV–12: *A grouped bar graph, for comparisons both within and between groups. Control data (0 Hz) are at the left in each group. Only the groups of bars are labeled. The bars are distinguished by diagonal line patterns, which are defined in a key. The patterns are simple and are easy to distinguish. Stripes are at a 45-degree angle and all run in the same direction. In addition, the patterns are symbolic: 0 Hz, no pattern; 5 Hz, light pattern; 10 Hz, dark pattern. There are spaces between the groups of bars but not between the bars within each group. All bars are the same width and the spaces between groups of bars and between the axis and the first group of bars are the same width. Bars are wider than the spaces. Line weights are as follows: boldest, outlines of bars; less bold, axis label and class labels; least bold, key, axis, tick marks, scale numbers, and statistical error bars. (Reproduced, with permission, from Johannsen UJ, Mark AL, Marcus ML. Circ Res 50:510, 1982.)*

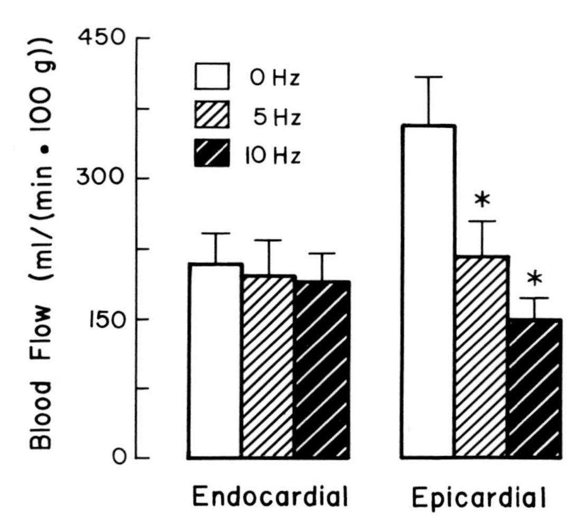

labels are generally aligned on the left (as in Fig. IV–11), but when class labels are of very different lengths, they are aligned on the right, along the baseline.

For groups of bars, individual bars can be labeled, and group labels can be centered below the individual bar labels (for vertical bar graphs) or placed to the left of the individual bar labels (for horizontal bar graphs). Alternatively, only the groups can be labeled and the bars can be distinguished by shading or patterns (as in Figs. IV–12, 13).

In bar graphs showing individual values, class labels are centered below each column of data points (Fig. IV–14), and group labels, if any, are centered below individual column labels (Fig. IV–15). Lettering in individual column labels is smaller than the lettering in the group labels.

FIG. IV–13: *A deviation bar graph. Groups of bars are labeled and individual bars are distinguished by shading. A key identifying the shading of individual bars is placed on the graph and extends only slightly beyond the rectangle implied by the axis and the baseline. (Reproduced, with permission, from Koizumi K, Terui N, Kollai M, Brooks CMcC. Proc Natl Acad Sci USA 79:2116, 1982.)*

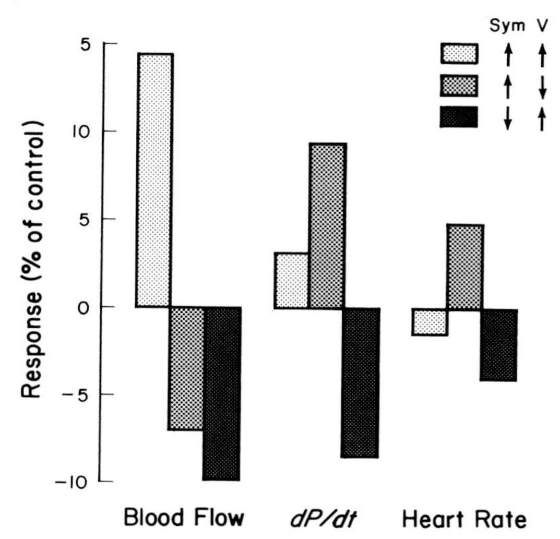

FIG. IV–14: *A variation on bar graphs in which individual values are shown by data points. The means are indicated by short horizontal lines. The baseline is drawn to identify zero. Class labels are centered below each column of data points. Data points are all black circles of the same size. When more than one symbol appears at one value, the symbols are placed in a horizontal row. When values overlap, the symbols are staggered. (Reproduced, with permission, from Donabedian H, Gallin JI. J Clin Invest 69:1155, 1982.)*

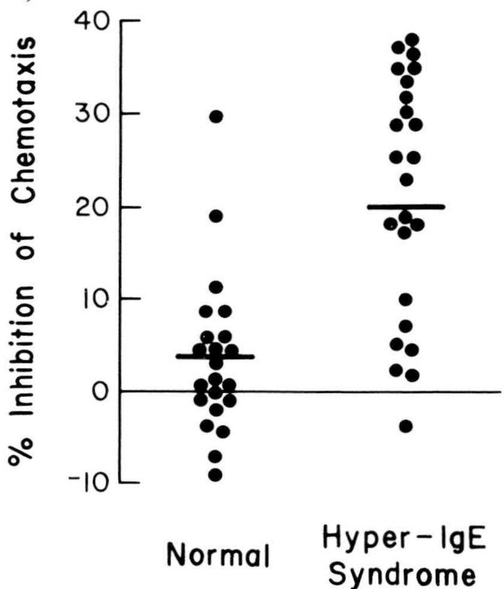

FIG. IV–15: *A variation on bar graphs in which individual values are shown by data points and the direction of the change is shown by lines joining the pairs of data points. For each column of data points, the mean and the standard deviation are shown by a data point and an error bar placed to one side of the column. Class labels are centered below the columns of data points. Group labels are centered below the class labels. (Reproduced, with permission, from Nolop KB, Braude S, Taylor KM, Royston D. J Appl Physiol 62:1244, 1987.)*

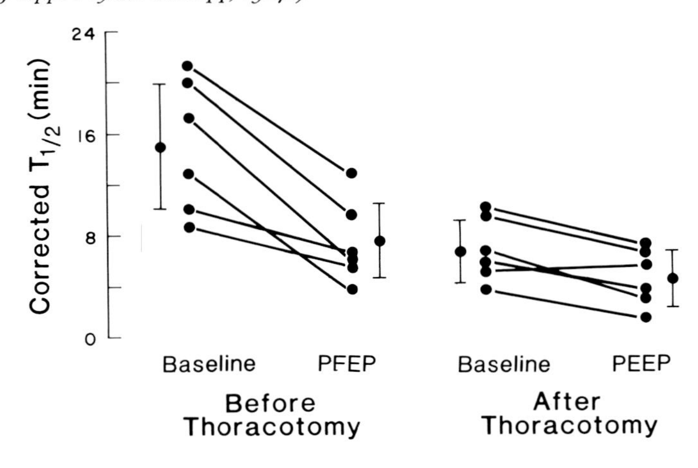

Class labels should be brief, but they can be longer in horizontal bar graphs.

Distinguishing bars

Bars can be distinguished by shading (fine dot screens) or by diagonal line patterns.

For shading, adhesive fine dot screens are available in seven gray values plus black and white (white, 0%; grays, 10%, 20%, 30%, 40%, 50%, 60%, 70%; black, 100%). A contrast of at least 30% for adjacent bars is wise, even if the line patterns differ, to ensure that the bars will be distinguishable after being reduced for publication (Figs. IV–12, 13).

A wide variety of adhesive patterns is available. Only diagonal line (striped) patterns that are simple and easily distinguishable should be used in bar graphs (see Fig. IV–12). Patterns that are excessively coarse, excessively fine, or distracting should not be used. Coarse patterns call attention to themselves. Fine patterns blur and are indistinguishable after reduction. Striped patterns should be at a 45-degree angle to the baseline and should all run in the same direction, to avoid optical illusions.

Like symbols used for data points in line graphs, shading and patterns can be used not merely to distinguish bars but also to group data (see Figs. IV–12, 13).

Whether or not shading or patterns are used, bars should have a bold outline so that they stand out from the background (see Fig. IV–12). A bold outline is particularly important for white bars. If bars do not need to be distinguished, they should all be black or dark gray (Figs. IV–10, 11).

Arranging bars and spaces

Bars should be arranged in a systematic order according to the experimental design. If baseline or control data are shown, they are conventionally placed at the far left in vertical bar graphs (as in Fig. IV–10) and at the top in horizontal bar graphs.

The longest bar (or its error bar) should reach nearly to the end of the graph.

All bars should be of equal width, and all spaces, including the space between the axis and the first bar, should be of equal width (Figs. IV–10, 11). Bars should be wider than the spaces betwen them. The exact amount of space depends on the number and width of the bars.

Bars should be neither very thin nor very wide.

When groups of bars are presented, the groups should be separated by a space, but no space is necessary between the bars within each group (see Figs. IV–12, 13). The space between the groups of bars should be about the width of one bar or less.

Showing individual values

For data points in individual value bar graphs that show values for a single population, black circles, all the same size, should be used. When more than one value occurs at one point, the symbols should be placed in a horizontal row and the row should be centered (see Fig. IV–14). When values overlap, the symbols can be staggered slightly off center so that a whole symbol is visible for each individual value (see Fig. IV–14). However, when columns of data points are joined, overlapping values should be drawn overlapped within the column (as in Fig. IV–15).

Lines showing means or connecting data points should be about twice as bold as the axis (Figs. IV–14, 15). Symbols for data points should be two to three times as wide as the lines showing the means or connecting the data points. The connecting lines should not touch black data points; thus, the shape of the data points will be clear after the graph is reduced to publication size.

The spaces between individual columns of data should be wider than the space between the axis and the first column (see Fig. IV–14). For columns of data points joined by lines, the joined columns should be wider than the spaces between the sets of joined columns (see Fig. IV–15).

The face of the graph

The actual amounts or frequencies should not be written on a bar graph, either within the bars or outside the bars. If exact values are important, the data should be presented as a table, not as a bar graph.

The virtue of bar graphs is their simplicity, so bar graphs should be as simple and uncluttered as possible.

Histograms and Frequency Polygons

DEFINITIONS

Histograms and frequency polygons are graphs that display a frequency distribution on a coordinate grid. A frequency distribution is a display that shows the total number of observations for each class of a ratio-scale variable, such as time or weight.

In a histogram, the frequency distribution is represented by a series of contiguous rectangles (Fig. IV–16). The width of each rectangle represents a range of values, which constitutes one class of the ratio-scale variable shown on the X axis. The height represents the corresponding frequency, which is shown on the Y axis. The area enclosed by the histogram represents the distribution of the data. Only if the widths of all

FIG. IV–16: *Two histograms. Each rectangle represents one class of a ratio-scale variable. The height of each rectangle shows the frequency for its class. Since the widths of the rectangles are equal, both the heights of the rectangles and the area of the histogram represent the frequency distribution. Histograms cannot be superimposed successfully; presenting histograms on separate axes, as done here, ensures readability. Note that the Y axes begin at zero. Also note that the A and B identifying the two parts of this composite figure are the largest and boldest letters on the graphs. For the composite figure, only one Y-axis label and one X-axis label are needed; also, only one set of scale numbers is needed for the X axis. (Reproduced, with permission, from Winslow RM, Monge CC, Statham NJ, Gibson CG, Charache S, Whittembury J, Moran O, Berger RL. J Appl Physiol 51:1411, 1981.)*

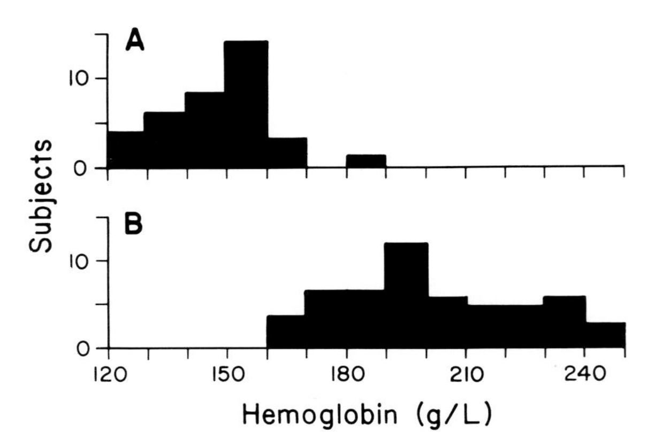

rectangles are equal do the heights of all rectangles represent the distribution. A histogram may be presented in outline only, with the vertical lines of individual rectangles omitted to emphasize the shape of the distribution (as in Fig. IV–16) or the vertical lines may be retained. The histogram may be shaded dark gray or black (as in Fig. IV–16) or left white.

In a frequency polygon, the frequency of each class of a ratio-scale variable is plotted at its midpoint and the midpoints are joined by lines (Fig. IV–17). So that the entire distribution will be represented, the lines joining the midpoints are extended to reach the baseline (zero frequency) at the midpoints of the classes that lie before the first class and after the last class. The area enclosed by the frequency polygon represents the distribution. Frequency polygons are useful for comparing overlapping distributions (see Fig. IV–17); this cannot be done successfully with histograms.

The age-and-sex pyramid comprises two histograms rotated so that the classes of the variable, age, are on the vertical axis (Fig. IV–18).

FIG. IV–17: *Two frequency polygons superimposed for comparison of overlapping distributions. Each data point represents the frequency of one class. The area of each frequency polygon represents one frequency distribution. Note that the Y axis begins at zero and that the lines joining the data points are extended to the baseline (zero frequency) at the beginning and end of each distribution. Note also the arrows placed above the data to identify points of special interest on the curve (in this case, standards to identify the void volume [^{125}I-LDL] and the lipoprotein-free fraction [^{125}I-albumin]). (Reproduced, with permission, from DeLamatre JG, Hoffmeier CA, Lacko AG, Roheim PS. J Lipid Res 24:1578, 1983.)*

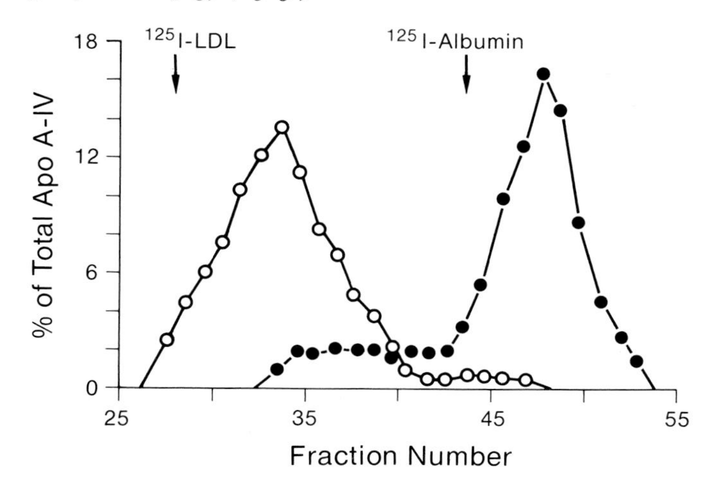

FIG. IV–18: *An age-and-sex pyramid in which the distributions of males and females are compared. (Reproduced from Shryock HS, et al.: The methods and materials of demography. Washington, DC, US Government Printing Office, Vol. 1, p. 242, 2ed., 1973.)*

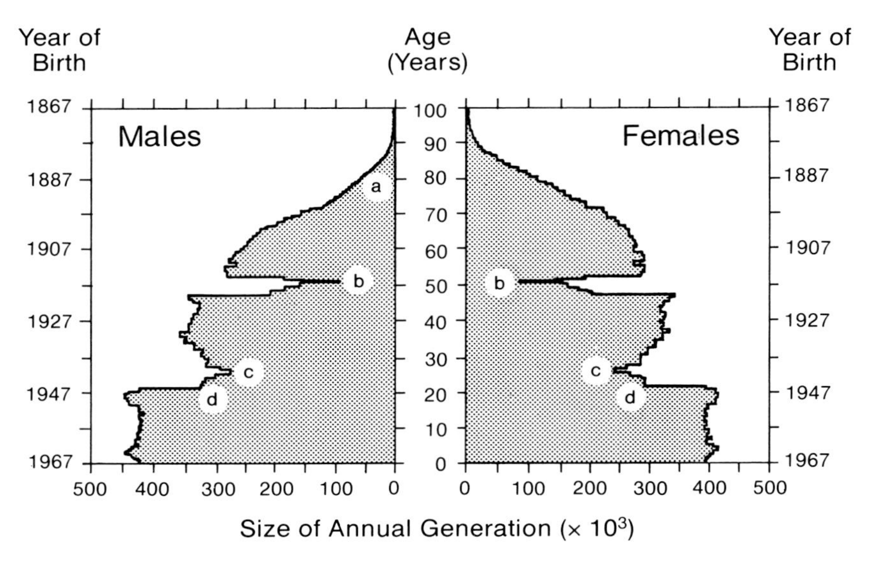

GUIDELINES FOR HISTOGRAMS AND FREQUENCY POLYGONS

The axes

The Y axis of histograms and frequency polygons, like the axis of bar graphs, must begin at 0 and must not be broken, so that the distribution is not distorted. The X axis does not need to begin at 0; the scale depends on the variable. Tick marks along the X axis of a histogram should be placed at the limits of each class (Fig. IV–16), not at the center of each class. For a frequency polygon, tick marks for the X axis are placed the same as on the X axis of a line graph (Fig. IV–17). Tick marks on both the X and Y axes should point outward so that they do not interfere with the histogram or frequency polygon.

The classes

In a histogram, classes of the variable shown along the X axis should be equal. If the classes are equal, the widths of all rectangles will be equal and the height of each rectangle will accurately represent the frequency of its class (as in Fig. IV–16). If the classes and the resultant rectangle widths are unequal, then the author must calculate the height of each rectangle so that its area will accurately represent the frequency of the class. Unequal classes make the histogram difficult to interpret, so they should be avoided.

In a frequency polygon, classes of the variable shown along the X axis must be equal.

Class intervals in histograms and frequency polygons should be neither very wide nor very narrow or the pattern of the distribution will be obscured.

Component bar graphs and pie graphs

DEFINITIONS

Component bar graphs (component bar charts) and pie graphs (pie charts) are diagrams that show the components of a whole. A component bar graph shows components by means of variously shaded or patterned segments of a bar (Fig. IV–19). A pie graph shows components as wedges of a circle (Fig. IV–20). For comparison of components of two or more wholes, component bar graphs should be used (as in Fig. IV–19) rather than pie graphs because the angular areas of the wedges of a pie graph are difficult to compare.

FIG. IV–19: *Component bar graphs. Components are distinguished by shading and patterns, which are identified by labels at the right of the graph. Component bar graphs rather than pie graphs are used because components of adjacent component bar graphs are easier to compare than are components of adjacent pie graphs. (Reproduced, with permission, from Shapiro CM, Bortz R, Mitchell D, Bartel P, Jooste P. Science 214:1253, 1981.)*

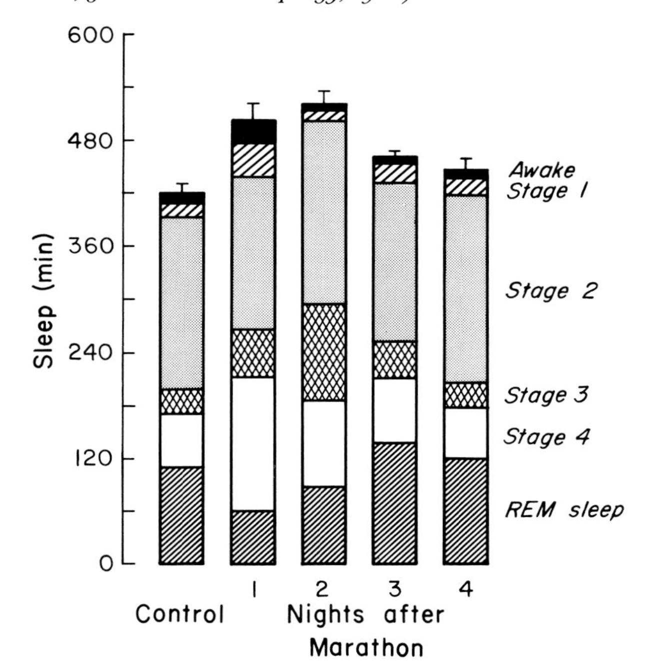

FIG. IV–20: *A pie graph. The percentage is given for each segment; the percentages total 100. Percentages are placed near the edge of the pie. Labels are separated from the percentages and are centered with reference to the percentage in the remaining space. For smaller segments, only percentages are placed inside the segment, near the edge, and labels are aligned in a column along the side. Leader lines run from each label to the appropriate segment. For the smallest segments, both the label and the percentage are placed in the column along the side. All labels and percentages are oriented horizontally. (Reproduced, with permission, from Clements JA. Composition and properties of pulmonary surfactant. In, Respiratory distress syndrome (Villee CA, Villee DB, Zuckerman J, eds), New York, Academic Press, pp. 77–95, 1973.)*

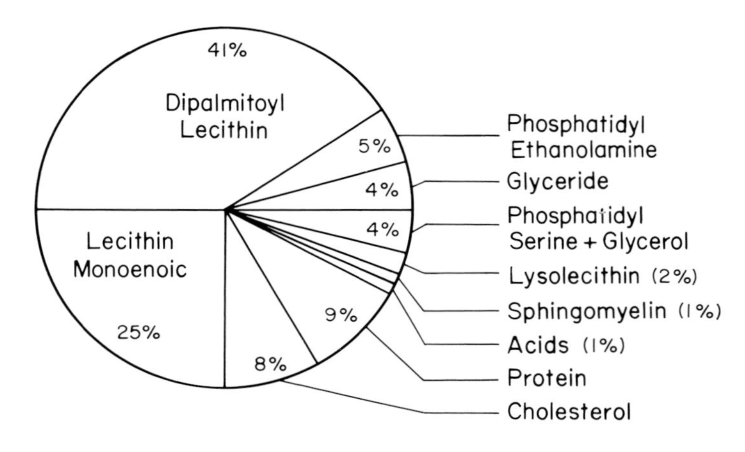

GUIDELINES FOR COMPONENT BAR GRAPHS AND PIE GRAPHS

General guidelines for constructing component bar graphs are the same as those for constructing other bar graphs.

Segments

In a component bar graph or a pie graph, the number of segments should be limited to those that can be seen and labeled. Components that are too small to be shown individually can be grouped into one segment labeled "other."

If the components are shown as percentages instead of as absolute amounts, each graph must total 100%. To convert from percentages to degrees, multiply the percentage by 3.6 (360°/100% = 3.6).

In a pie graph, the largest segment conventionally begins at 12:00 or at a quarter hour and runs clockwise; remaining segments continue clockwise. The segments are organized according to experimental design or according to size.

Guidelines for shading and patterns in component bar graphs and pie graphs are the same as those for distinguishing bars in bar graphs. Segments of pie graphs may be left white.

Segment labels

In a component bar graph, labels are not placed inside the segments because they clutter the graph unduly. Instead, labels are placed to the right of a vertical graph (Fig. IV–19) or above a horizontal graph. Alternatively, segments are identified by a key.

In a pie graph, labels are placed inside the segments if the labels are brief and the segments are large enough. Otherwise, labels are listed in a column next to the pie graph (or two columns, one on each side of the graph) and leader lines are run from each label to the appropriate segment. A combination of the two systems can also be used: labels inside large segments, and labels in a column for small segments (as in Fig. IV–20). Because it is difficult to estimate angular areas, the percentage should be included in each segment or after each label of a pie graph.

The labels and percentages in the segments and the column are placed horizontally, not at an angle. Percentages are placed near the edge of the pie. Labels are separated from the percentages and are centered with reference to the percentage in the remaining space. Numbers are slightly smaller than the capital letters in the label. Leader lines touch the outline of the pie graph but do not touch the letters of the labels. Leader lines are less bold than labels and segment lines.

Statistical Information

ERROR BARS AND NUMBER OF OBSERVATIONS

In line graphs and bar graphs, variability surrounding the mean value can be indicated by an error bar (a thin vertical line ending in short horizontal lines) centered at each data point (Fig. IV–6) or data bar (Fig. IV–10). The figure legend should state whether the error bars represent standard deviations, standard errors of the mean, confidence limits, or ranges. The number of observations (n) should also be stated, preferably in the legend. Putting n next to each data point or inside each data bar clutters the graph.

Generally, full error bars can be drawn (as in Fig. IV–6). However, in line graphs, if error bars create a lot of clutter, half error bars can be drawn (see Fig. IV–6, "old" curve, three lowest data points). Half error bars can also be drawn in bar graphs that have gray or black bars (Figs. IV–10–12).

When error bars are not appreciably larger than data points or data bars, error bars can be omitted. The omission should be mentioned in the legend.

FIG. IV–21: *A scattergram in which 95% confidence limits of the regression line are shown by dashed curved lines. (Reproduced, with permission, from Soifer SJ, Morin III FC, Heymann MA. J Pediatrics 100:458, 1982.)*

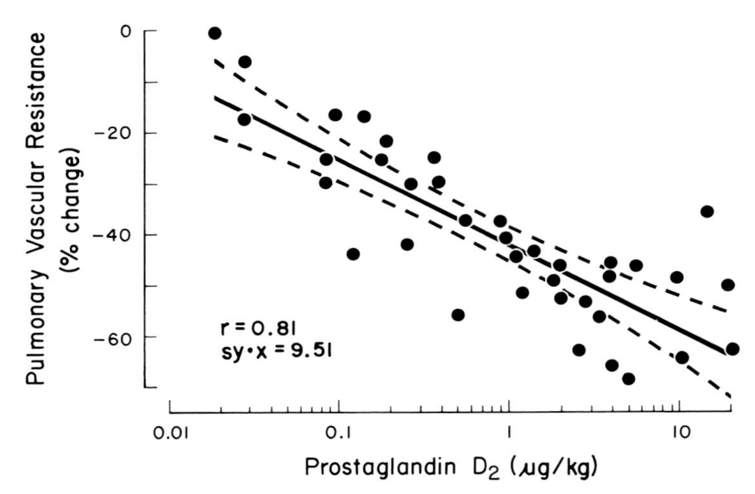

CONFIDENCE LIMITS

Confidence limits for regression lines or for populations can be shown by dashed or solid lines (Fig. IV–21). Note that confidence limits are slightly curved.

SIGNIFICANT DIFFERENCES

In a bar graph, statistically significant differences between bars can be indicated by a symbol (usually *) placed at the end of the data bar that is different (Fig. IV–12). The symbol is not placed above the control bar or between data bars. If error bars are included, the symbol can come either at the end of the error bar (Fig. IV–12) or above and slightly to one side of the data bar.

Probability values (*P* values) should not be put on the graph either instead of or in addition to symbols indicating statistically significant differences, because *P* values clutter the graph. The *P* value, preceded by a statement telling which data are being compared, belongs in the legend. For example, "*, significantly different from the control value, $P < 0.01$."

For more than one statistically significant difference, one commonly used series of symbols is *, $P < 0.05$, **, $P < 0.01$, ***, $P < 0.001$. However, these lengthy statistical symbols can clutter a graph. It is best to keep the number of different statistical symbols to a minimum. If necessary, specific *P* values can be grouped; for example, $P = 0.015$ and $P = 0.02$ can be grouped as $P \leqslant 0.02$.

Brackets or other lines should not be used to indicate which data are being compared, because they clutter the graph.

Differences that are not statistically significant do not need to be identified: the presence of a statistical symbol identifies differences; the absence of a statistical symbol on the same graph implies no significant difference.

Sets and Composites

All the graphs for one paper should be designed as a set. Each graph prepares the reader's eye for the next one, so the graphs should be as similar as possible in scale, dimensions, and style. For example, when the same dependent variable is shown in more than one graph in a manuscript, the same symbol should be used to represent it. In addition, the Y-axis labels for all graphs should have the same orientation. This notion of a set can be extended to include illustrations of various types, for example, a line graph showing mean electrocardiographic data and an individual electrocardiographic recording.

Related graphs can be combined into a composite illustration, in which two or more graphs appear side by side or one above the other (see Fig. IV–16), or both. The composite, along with its legend, should be designed to fit across the journal page or within a column.

In a composite figure, redundant axis labels and redundant scale numbers should be omitted, as in Fig. IV–16.

The graphs within a composite should be of the same proportions and style. Graphs in a composite that are to be compared should also have the same scales. When the same dependent variable appears in more than one graph of a composite, it should be represented by the same symbol, just as in a set.

Each graph within a composite should be identified either by a brief label naming the distinguishing aspect or by a capital letter (Fig. IV–16). The labels or letters should be placed as nearly as possible in the same spot in each graph, preferably in the upper left corner.

Legends

The legend is a descriptive or informative statement, set in type, that appears below or next to an illustration in a published article to identify and explain the illustration. The legend is needed so that the illustration will be intelligible without reference to the text.

The legend of a graph typically has four parts: (1) a brief title, which states the topic or the point (purpose, message) of the graph; (2) one or more sentences giving experimental details; (3) definitions of symbols or abbreviations not explained earlier in the legend; and (4) explanations of statistical comparisons.

TABLE IV–1
Line and Letter Weights in Graphs

Weight	Type of Line or Letter				
	Line graph*	Bar graph, Component bar graph	Individual value bar graph	Histogram, Frequency polygon	Pie graph
Boldest†	Curves	Outline of bars	Lines showing means or connecting data points	Outline of graph	Outline of graph
Less bold	Axis labels Outlines of white data points Confidence limits	Axis and class labels	Axis labels	Axis labels	Segment lines Segment labels
Least bold	Axes Tick marks Scale numbers Curve labels Keys Statistical error bars	Axis Tick marks Scale numbers Keys Segment labels‡ Statistical error bars	Axes Tick marks Scale numbers Baseline (if drawn) Statistical error bars	Axes Tick marks Scale numbers	Leader lines

* For insets in line graphs, line weights are the same as for the main graph, except that curves are less bold and letters and numbers are smaller. Data points are also smaller.
† In composite figures, the letters or words that identify each part of the composite should be the boldest letters on the graph.
‡ For component bar graphs.

Line and Letter Weights, Sizes, and Styles

To ensure that the information on graphs stands out, make line and letter weights follow the importance of the information. That is, make the curves and data points or data bars the most prominent features. The curves and outlines of data bars should be bolder than the axis and class labels, and the labels should be bolder than the axes (see Table IV–1). The curves should not be so bold as to be overwhelming.

Similarly, make letter sizes follow the importance of the information. Axis labels and class labels should be the largest letters on the graph. Curve labels and keys should be 75%–90% of the size of axis and class labels. Scale numbers should be 75%–90% of the size of the capital letters in the axis and class labels.

In a composite figure, the letters or labels that identify each part of the composite should be the largest and boldest letters on the figure, about 125% the height of a capital letter in the axis labels and class labels (as in Fig. IV–16).

For maximum readability, labels should be in upper and lower case letters in a vertical (not slanted), uncrowded, sans serif type face of medium weight (see Chapter III). The combination of upper case (capital letters) and lower case is easier to read than all upper case letters and also is less likely to overwhelm the data. Vertical lettering is used for all labels, though slanted (italic) lettering may be used for curve labels or keys (as in Figs. IV–5, 7, 19).

Units of measurement, usually International System (SI) abbreviations, should be given in parentheses in lower case letters, unless the abbreviation is conventionally upper case (for example, °C).

To check that symbols and lines are neither too small and thin nor too large and bold, use a reducing lens, or reduce the sketch on a reducing copier until it reaches the size that it will appear in the journal.

Artwork

Graphs are easiest to draw two to three times larger than they will appear in print. Start by measuring the width of one journal column (or more, depending on the complexity of the graph). Multiply this by 2 to 3 (200%–300%) to obtain the width to prepare the original artwork (see Chapter III). A one-column graph can usually be drawn on standard size paper, such as 8½ × 11 inches (216–279 mm) or A4 paper (212 × 297 mm). The final x-height of the smallest lower case letters should be no less than 8 point type (1.5 mm). For reduction to 50%, then, the original height of lower case letters will be no less than 3.0 mm.

Multi-purpose Graphs

Graphs prepared for publication are not ideal for 35-mm slides or posters. Readers can study published graphs from whatever distance and for however long they like. Therefore, published graphs can have more content and smaller letters. Readers of slides, on the other hand, do not control reading distance or time. Nor does the slide have an explanatory legend. Therefore, the content of slides must be more limited and the letters must be larger. In addition, a title should be placed on the slide (centered at the top). Graphs for posters fall somewhere between graphs for publication and graphs for slides. Original artwork is generally enlarged for posters—usually to 200% or 300% of the original—so that words can be read from about 2 meters. Thus it is important to consider all possible uses before preparing a graph.

Color

For virtually all purposes graphs can be drawn in black on a white background. Black on white is clear and is easy to follow. Color is rarely necessary or desirable in graphs. Also, color is difficult to design effectively; if used ineptly, color detracts from the message of the graph. Finally, color is an unnecessary added expense.

Computer-produced Graphs

Graphs produced by computer, to be acceptable for publication, must adhere to the same standards as graphs that have been prepared manually. Increasingly, this is becoming possible (see Chapter V).

Advantages of computer graphing include speed and the ability to manipulate scale and to explore alternative ways of presenting data. Computers are particularly useful for displaying data in three dimensions.

A major disadvantage is that the quality of graphs produced by many desktop computers does not yet meet publication standards. However, as development of computers continues, quality should improve (see Chapter v).

Maps

Maps are used not only to delineate and describe the earth's surface (geographic maps) but also to show relationships involving area, distance, or direction (distribution maps). Whereas tables focus on numerical values of data and graphs on relationships of data, distribution maps focus on the location of data. The data can be either quantitative — rates, ratios, amounts, frequencies, or distributions — or qualitative — location of races, religions, climatic conditions, or flora and fauna, for example, without regard to numerical value.

A successful map, whether prepared manually or by computer, requires technical and design skills, as well as insight into the problems of visual perception and readability. These skills and insight are best provided by professional cartographers or map drafters. In this chapter, we describe the types of distribution maps and present guidelines for designing each type.

Common Types and Purposes

For quantitative data, choropleth maps are used to show rates, ratios, or frequencies based on areas; point symbol maps are used to show frequencies or amounts; isoline maps are used to show distribution of absolute or derived values; flow line maps are used to show flow of data; proportional maps and choropleth maps are used to show rates, ratios, or frequencies based on areas.

For qualitative data, the point symbol map, the area (or zone) map, and the isoline map are used to represent qualities or attributes of a location.

Quantitative Maps

Choropleth maps

A choropleth (khoros, place; plethos, quantities) is a map that uses gray

FIG. IV–22: *A choropleth map showing number of hospitalized cholera patients in Portugal, April-October, 1974. Note that zero cases is represented by white.*

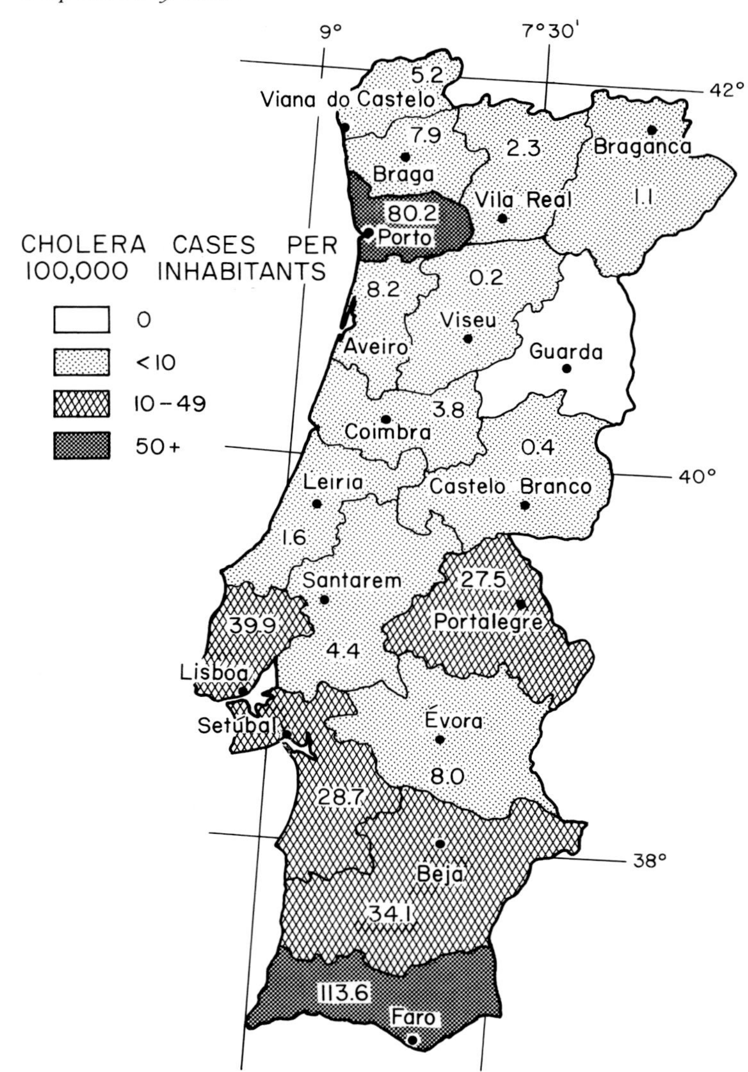

tones to represent rates, ratios, or frequencies within political or geographic areas (Fig. IV–22). The main use of the choropleth is for display of demographic data, for example, rates of birth and morbidity; population density and change; incidence of disease; sightings per square kilometer. The choropleth is especially useful when numerical values within regions are inconsistent.

Disadvantages of the choropleth arise from the selection of geographic areas. Large areas appear more important than small areas, regardless of their numerical value. Also, it is impossible to show subtle changes of value or blending from area to area or a range of values within the chosen area.

Choropleths are prepared by first organizing the data into an array as for a graph frequency distribution and then selecting a maximum of seven intervals. In a graph of frequency distribution, class intervals are normally of equal size, e.g., 0–4.9, 5.0–9.9, 10.0–14.9, 15.0–19.9, etc. Although this is also desirable on a choropleth map, it is not always practical. Instead, intervals should be chosen so that the number of political or geographic areas within each class is similar. Thus it is not uncommon to find a range such as 0–4, 5–14, 15–24, 25–49, 50–99, 100–299.

Gray tones from light to dark are chosen to correspond with the numerical values from low to high. Perceptual problems can arise because readers tend to overestimate differences between gray tones and thus overestimate differences between numerical values. Also, jet black areas become very important. Another problem is that the density of gray tones may alter on reduction. A key to values must be provided.

Point symbol maps

Point symbol maps are maps on which point symbols are used to represent absolute values or frequencies at specific locations. Dots, circles, spheres, or cubes can vary in number (multiple dot maps) or size (gradu-

FIG. IV–23: *A multiple dot map showing the amount of wheat harvested in the United States in 1969 (in acres).*

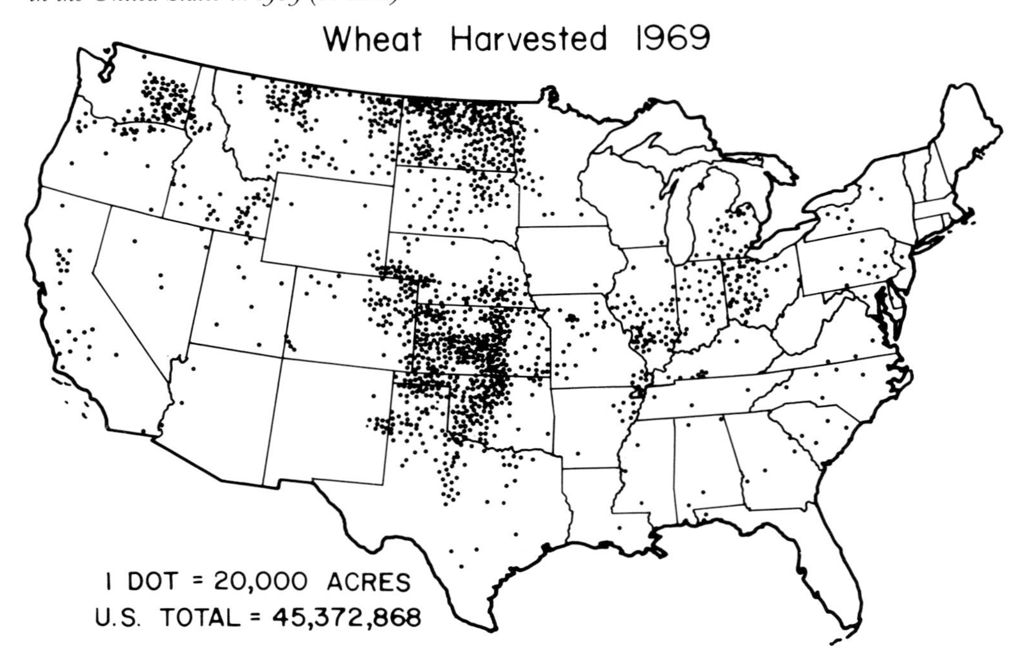

ated dot maps), depending on the type of data displayed and what is to be emphasized. Point symbol maps cannot be used to show ratio or relationship.

MULTIPLE DOT MAPS

In multiple dot maps, small dots of equal size, each representing one value, are used to show density or clustering (Fig. IV–23). Multiple dot maps are easy to prepare because little calculation is required. A common use is the display of population density. Other uses include showing the number and location of bird sightings or volcanic eruptions.

Dot size must be carefully chosen — if the dots are too large, many dots will overlap; if dots are too small, obvious patterns will remain unseen. The choice of the unit value that each dot represents is also of concern. Choosing too small a unit value results in too many dots; too large a unit value results in too few dots. When optimal dot size and unit value

FIG. IV–24: *A graduated dot map with proportional area scaling showing population density in the northeastern United States in 1970.*

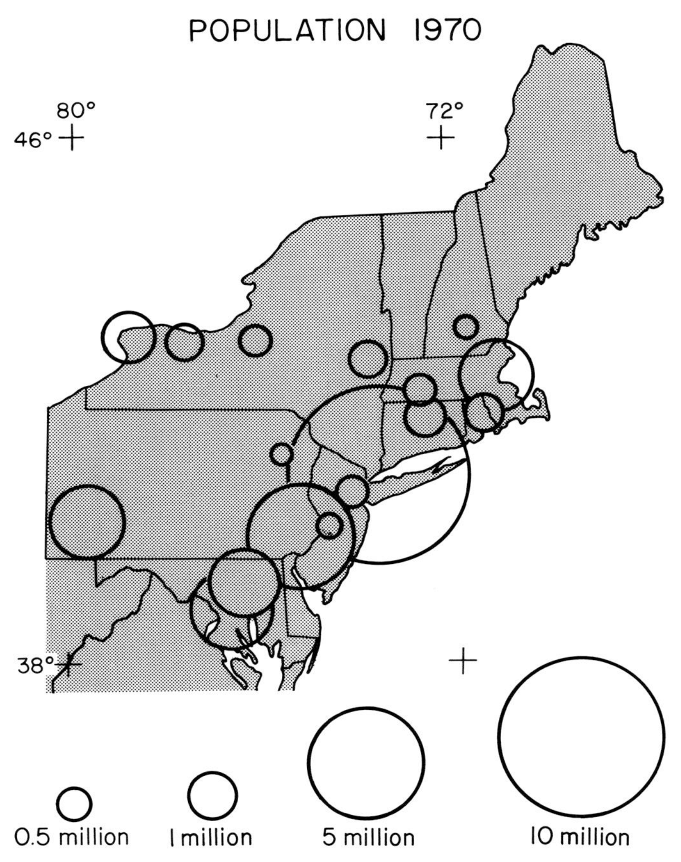

have been chosen, dots in the denser areas will just barely run together and patterns will be clearly visible. It is best to show only one kind of data, or two at the most, for then the visual impression is clearest.

Choosing different symbols to represent different amounts, for example, ○ = 0, ● = 10, is not advisable.

GRADUATED DOT MAPS

In graduated dot maps, dots of different sizes are used to represent different values (Fig. IV–24). Graduated dot maps are used to display a range of values or absolute amounts for areas or discrete locations when totals are more important than exact location. Examples include the display of total bird population for each country and quantity of coal mined by mine site.

The size of the symbols should be proportional in area to the quantities represented. For example, in Fig. IV–24, the area of the circle (πr^2) representing 15 million people is twice that of the circle representing 7.5 million people. Since comparison by area of graduated circles appears to minimize differences, whereas comparison by diameter greatly exaggerates them, symbol sizes are sometimes "psychologically" scaled with the use of correction formulas. A simpler way to minimize perceptual errors is to provide a scale of legend symbols that are not nested within each other (Fig. IV–24).

FIG. IV–25: *A map on which total regional wheat imports and exports are represented by bar graphs. (Data from the New York Times atlas of the world, New York, NY, Times Newspapers Ltd., rev. ed. 1980.)*

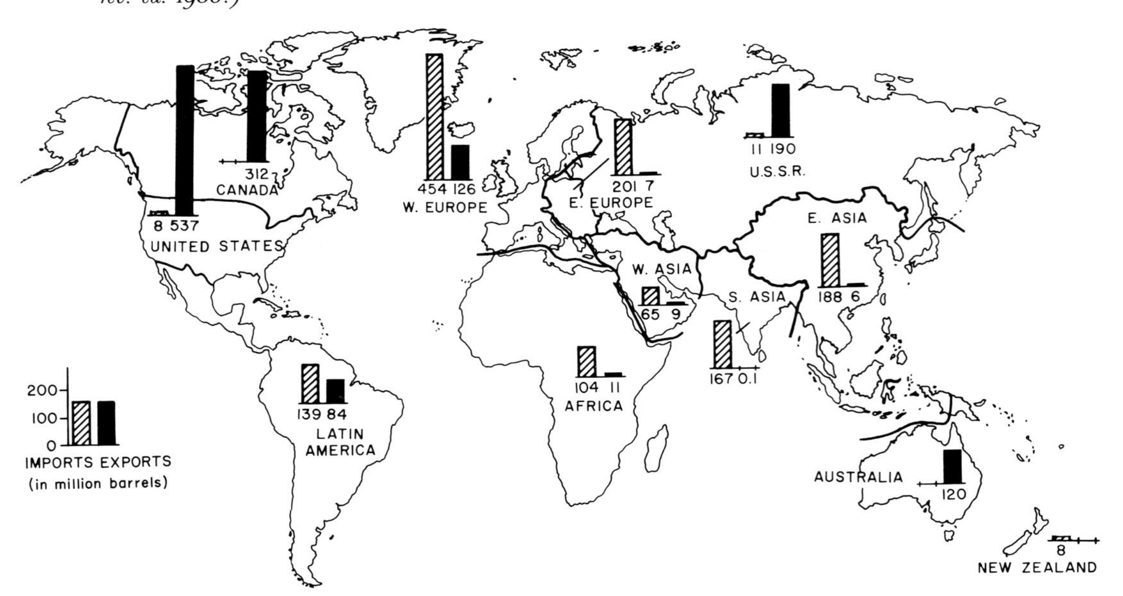

Another method of scaling graduated symbols is to classify the data into ranges of values. As in the choropleth, no more than seven classifications should be used; each range of values is assigned a symbol of differenct size. The area of these symbols, too, should be directly proportional to the quantities represented.

OTHER SYMBOLS USED ON POINT SYMBOL MAPS

Other symbols that can be used to show total amounts at a site or in an area are bar graphs, pie graphs, stacked squares, and graduated spheres.

Bar graphs, because they vary in only one dimension, are not difficult to compare, but since the baselines of the bars are not all situated on the same line, perceptual error can be a problem. This is minimized if the total values are included adjacent to the graphs (Fig. IV–25).

Pie graphs can be used to show percentage composition as well as totals (Fig. IV–26). The circles should all be subdivided from the same

FIG. IV–26: *A map on which both total values (total cereal production by region, in parentheses) and component values (proportion of total production comprised of individual grain crops) are represented by pie graphs. (Data from the New York Times atlas of the world, New York, NY, Times Newspapers Ltd., rev. ed. 1980.)*

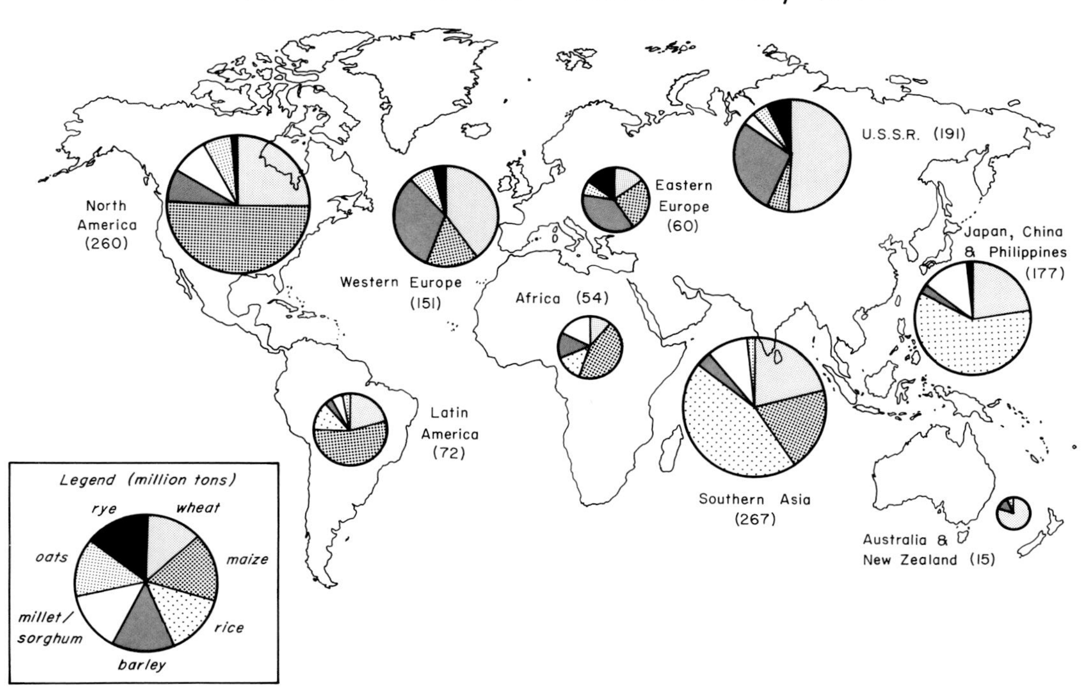

starting point to simplify comparisons. Accurate comparisons of the sizes of circles, however, are difficult.

Squares of equal size can be used effectively to show totals by being stacked into blocks. Different sizes of squares, however, present reading difficulty and are not recommended. For the same reason, spheres or cubes are not advised, although spheres can be used to display widely dispersed values.

Data can be located with actual numbers, but displaying numbers on a map is not as clear as displaying them in a table or a bar graph.

Isoline maps

In the isoline map, lines join points of equal value to indicate boundaries of various quantities (Fig. IV–27). This type of map is used to indicate

FIG. IV–27: *An isoline map relating the similarity of different floras (dots) with isoline values derived from a principal component analysis based on similarity indices. (Data from Steven P. McLaughlin, by permission.)*

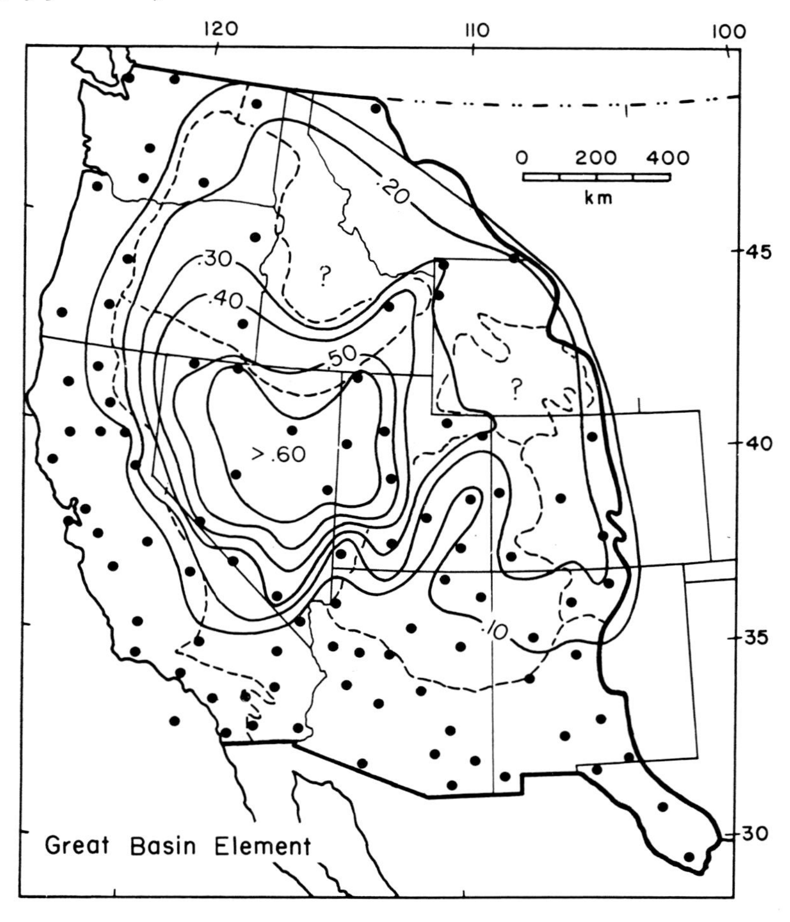

boundaries for such variables as temperature, elevation, or rainfall, as well as derived values (e.g., means, medians, or ratios) such as average monthly rainfall, ratio of sunny days to cloudy, or number of people per square kilometer.

Flow line maps

A flow line map is a map that uses arrows of different direction, length, and width to display the flow and amount of data in a geographic area (Fig. IV–28). Applications of flow line maps include waves of migration and flow of traffic and commodities. The thickness of the lines are directly proportional to the values being represented; that is, a 1 mm wide line might represent 100 million tons of oil exported, and a 10 mm wide line 1,000 million tons (see Fig. IV–28).

Proportional maps

A proportional map is a map that displays rates, ratios, or frequencies within political areas by means of distorting the area in proportion to

FIG. IV–28: *A flow line map showing oil shipments by sea of OPEC countries in 1976. (Data from the New York Times atlas of the world, New York, NY, Times Newspapers Ltd., rev. ed. 1980.)*

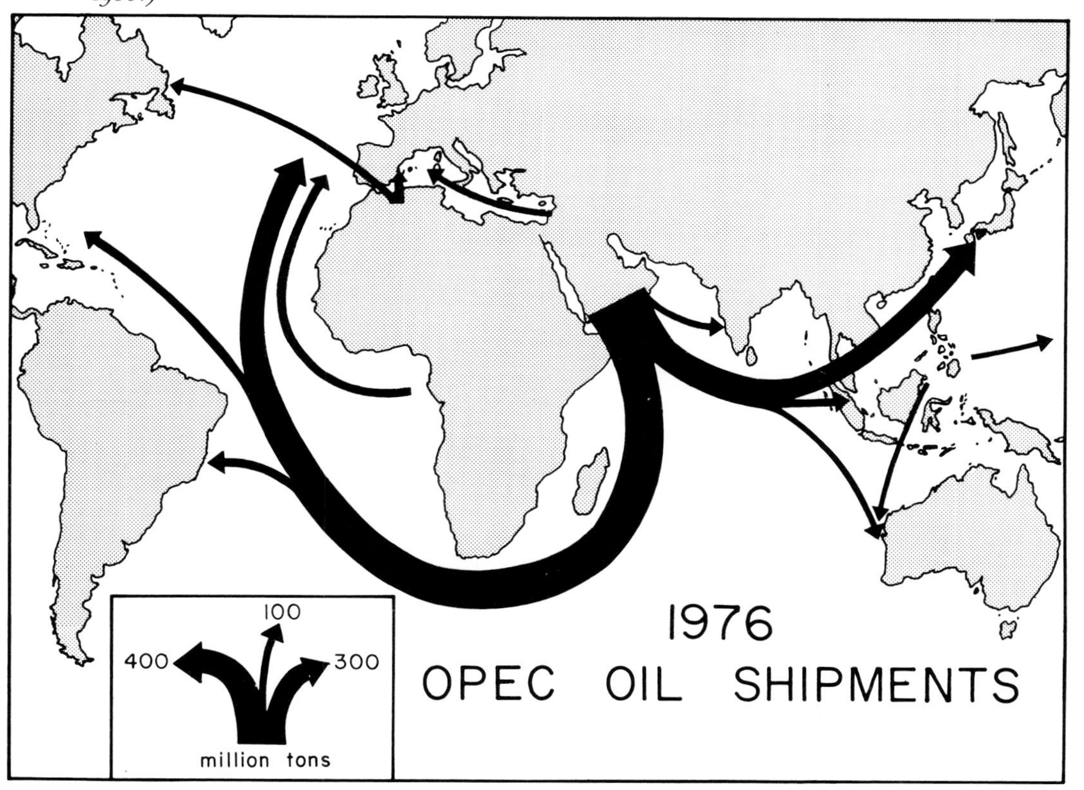

FIG. IV–29: *A proportional map showing the relative population of the states of the United States. The area of each state is proportional to its population.*

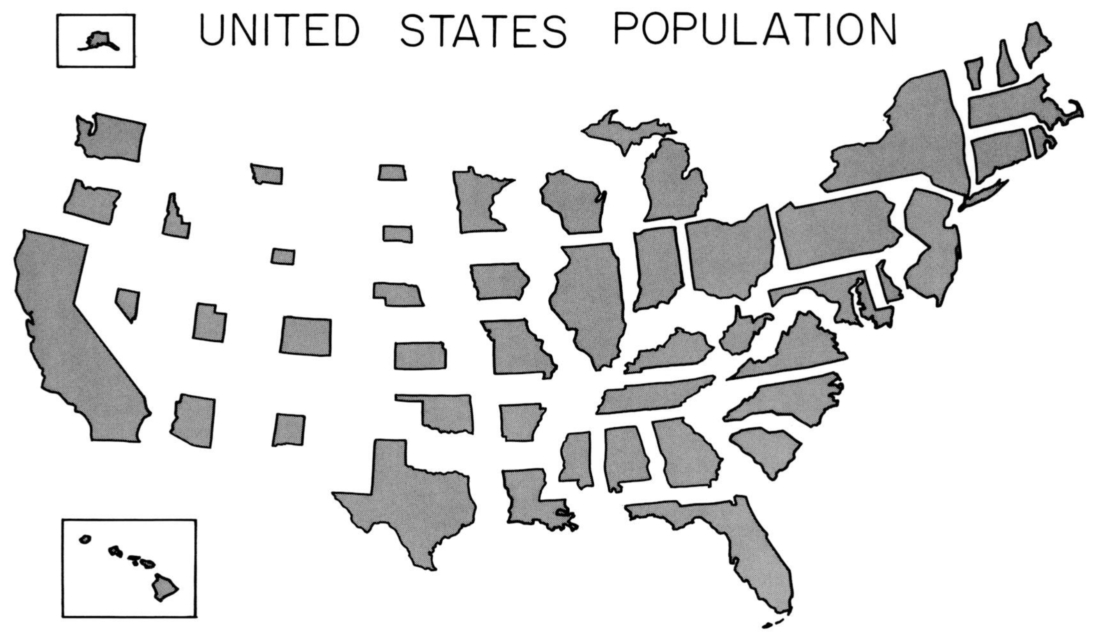

the numerical value of the data (Fig. IV–29). Thus, in a population map of the United States the eastern states would be drawn very large and the northwestern and southwestern states would be small. The trick is to draw the map so that the area of coverage is still recognizable. For this reason, the proportional map should be used only for well-known areas.

Qualitative Maps

Qualitative information can be shown with point symbols, delineated areas, or isolines.

Point symbol maps

On the point symbol map, different symbols (dots, triangles, squares) in different hues (black, white, colors) are used to show various attributes of an area without regard to quantity (Fig. IV–30). This type of map is used to show point locations of different types of plants or animals, for example. A maximum of five types of symbols is recom-

FIG. IV–30: *A point symbol map showing distribution of four species of* Mammillaria *in Arizona. (Data from Benson L: The cacti of Arizona. Tucson, AZ, University of Arizona Press, 1969.)*

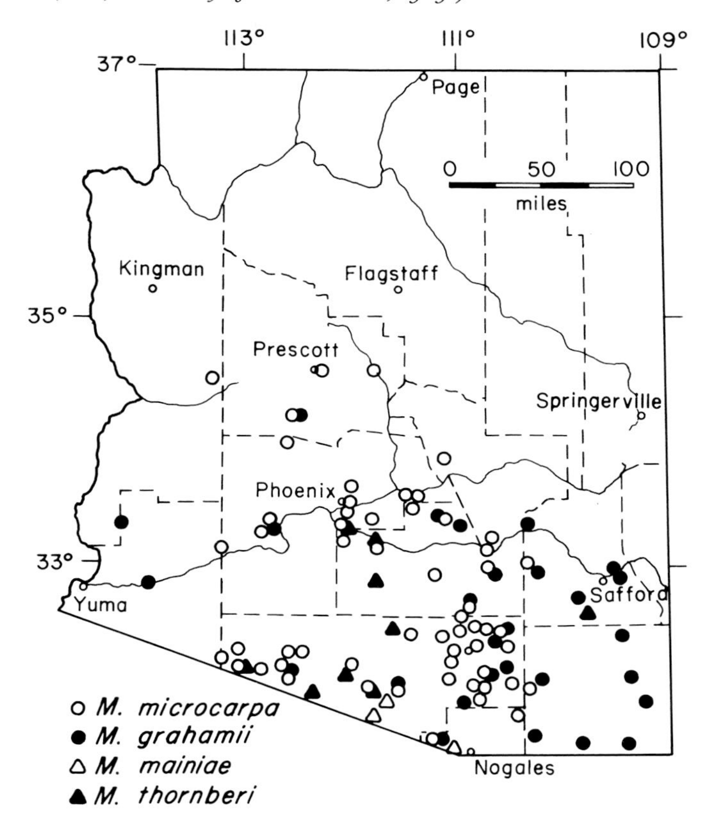

mended but this depends on the distribution and amount of data displayed. Choice of black and white symbols is discussed on pages IV–00–00.

Area maps

On the area or zone map, different gray tones or different colors are used to indicate the areas in which various attributes are located within a geographic region (Fig. IV–31). This type of map is used to illustrate distribution of different religions, racial groups, or taxonomic groups, for example, without regard to quantity.

Isoline maps

On qualitative maps, isolines are used to locate boundaries of things or events, such as epidemics, without regard to quantity (Fig. IV–32).

FIG. IV–31: *An area map showing distribution of species of the* Castilleja viscidula *group. Note the overlapping areas. See Fig.* VIII–1 *for use of spot color to indicate overlapping areas. (Data from Noel Holmgren.)*

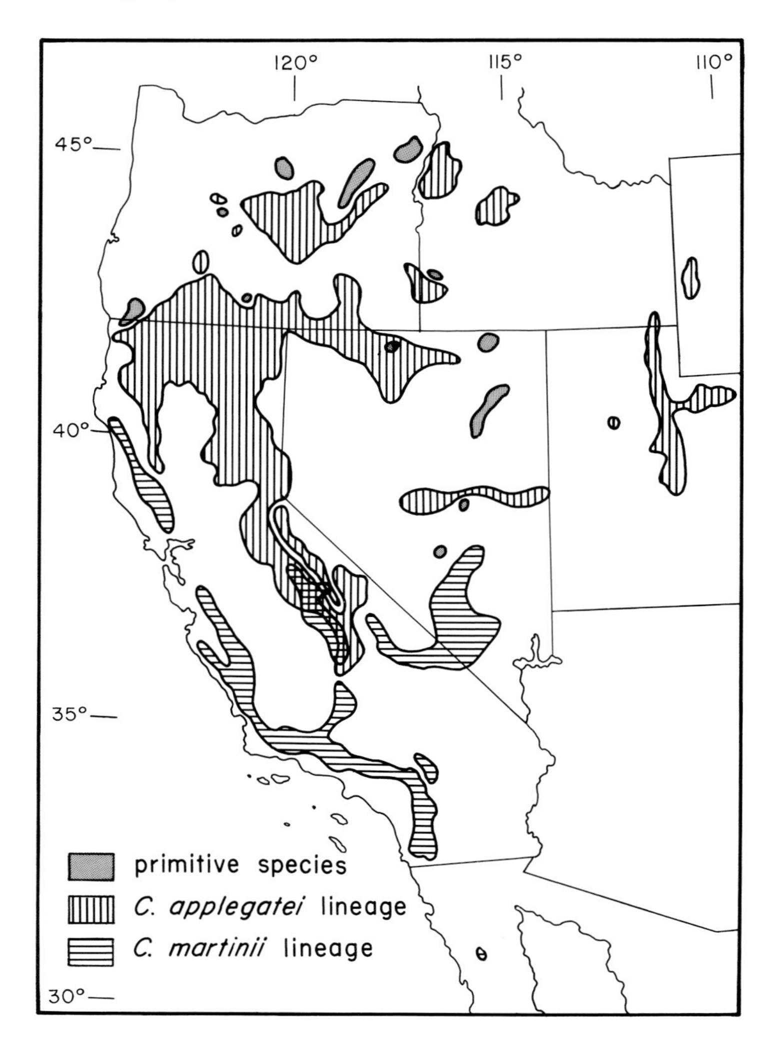

Guidelines for Preparing Maps

Message statement

The first step in designing a map is to determine the message. The message is a simple verbal summary of the purpose and content of the map. Failure to identify precisely what the map is to say often leads to inclusion of extraneous details that obscure the message. A statement of the message should be included in the legend or incorporated into the title.

FIG. IV–32: *An isoline map showing expansion of cholera during the sixth pandemic (1899–1922).*

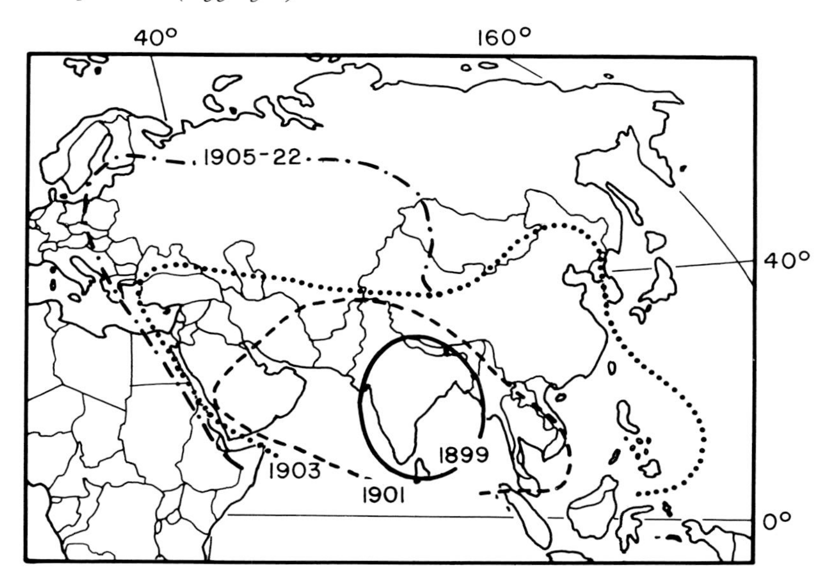

Base maps

The base map (an outline map with minimal labeling) provides the foundation on which to build and enables the display of information.

TYPES OF PROJECTION

Any two-dimensional rendering of the globe is a distortion. Many projections have been designed to reduce these inherent distortions; each has uses for which it is suited. Equal area projections, such as Mollweide's Homolographic and Sinusoidal projections, are suited to comparison of land areas. Conformal projections, such as Lambert's conic and Mercator's projections, retain directional integrity, yet distort areas. They are useful for navigation or meteorology maps.

COMPLEXITY OF RENDERING AND DETAIL

The use to which the base map will be put determines the amount of detail needed. The goal is to present only that information which will be used and to present it in hierarchies of visual importance. Background details should not compete with essential information. On reduction, all details should be clearly readable.

CHOICE OF SCALE

The choice of scale is largely determined by the map's purpose. Maps that are studied intensively or are used to record and store or to measure and analyze information will require great detail and accuracy and a large scale. Maps on which large areas are to be compared should have a small scale.

SIZE

Choice of base map size depends on its use and also on the final printed size. Maps are generally drawn larger and are reduced to 50% of the original. It is therefore essential that all words and details are readable at this reduction. A drafter, using a graphic arts projector, can reduce or enlarge the map to suit the scientist's needs.

WHERE TO FIND BASE MAPS

University map libraries are often the best sources of base maps. A knowledgeable map librarian can direct an author to other specialized collections. Atlases, textbooks, or public agencies are also sources. Because published maps may be copyrighted, permission to use a map must be obtained and acknowledged. United States Government maps and maps from the U.S. Geological Survey are not copyrighted, and therefore may be used without obtaining permission. Sources: National Cartographic Information Center, U.S. Geological Survey; U.S. Census Bureau Data Users Services Division; university cartography departments.

MAP SERIES

The base map is well suited to production of a series of maps of the same subject area. The base map can be duplicated and then each copy rendered differently, or it can be used as a base for a series of overlays. In this way, the temptation to overload a map is reduced and all subsequent maps are of a uniform size.

Lettering

Type, size, and placement of lettering all act as symbols and must therefore be chosen carefully. Words are used to locate specific geographical points, such as cities. If stretched out, words can indicate orientation and length of linear phenomena, such as rivers, or can designate form and extent of geographic areas. Size and style can classify and differentiate various features. For example, all hydrographic features would have one style, all cities would have another. The author should present to the cartographer a list of words arranged in hierarchies of importance so that, by varying size, style, or boldness of type, the cartographer can create the appropriate contrasts or draw attention to important areas.

Care should be taken to ensure that all wording is readable on final reduction. The x-height of the smallest order of lower case letters should be no less than 1.5 mm. For a planned reduction to 50%, for example, the original height of the smallest lower case letters should be no less than 3.0 mm. Generally speaking, letters should be larger than this.

Positioning of the type is as important as size and style. Incongruous, sloppy positioning is readily apparent to the reader. Ideally, names should be either entirely on land or entirely on water and should be oriented parallel to the upper or lower edges of the map and right-side up. Lettering should not be curved except to identify curvilinear features such as rivers or boundaries. Where lines conflict with lettering, the lines, not the names, should be interrupted. In order that words over hatched areas will be readable, a narrow border of white should be left (Fig. IV–22).

Gray tones

Gray tones are used extensively in the choropleth and are generally selected to correspond to numerical values of the chosen ranges. Light grays represent lower values and dark grays higher values, regardless of the patterns used. All areas having a numerical value must have a gray tone. White is used only to depict zero values or the absence of data (as in Fig. IV–22). Seven gray tones are the maximum that can be easily distinguished. Patterns that will reduce well should be chosen. Fine patterns may block up or disappear on reduction.

Commercially prepared self-adhesive patterns are available. These simplify the job of hatching large areas with pen and ruler, but may not stick well to certain ground materials or may not roll well for map storage or mailing.

Another method of preparing maps with different gray tone patterns is to use a separate overlay for each gray tone. An opaque material such as Rubylith is cut to fill the area. On each overlay, instructions about color or percent gray tone are noted. The printer then photographs each overlay to prepare the final negative and plate for printing. This method, carefully done, generally results in a better quality product (see Chapter III, "overlays," for details).

Color

Although color is seldom used in statistical maps because of the cost of reproduction, it has distinct advantages over gray tones. In addition to its visual appeal, color has a much wider range of hues and values. These can be used to identify more features in more meaningful ways.

In quantitative maps, various intensities of one color should be chosen, for example, light blue to dark blue, or pale yellow to orange to brown, to represent increasing numerical values. In qualitative maps, on the other hand, a variety of colors can be used. Unless one area is more important than others, no single color should appear more important than any other color.

Lines

Lines that are too fine may disappear on reduction. Good black ink should always be used, either on brilliant white paper or on drafting film.

Latitude and longitude

In maps such as route maps or maps of small areas that may not be easily located geographically, latitude and longitude must be identified, although a complete grid is seldom necessary. Geographical coordinates can be placed unobtrusively at the edges of the map (as in Figs. IV–22, 24).

Title, legend

Maps that must stand alone will have the message statement in the title. Maps used as figures to illustrate the text can be less cluttered if the message statement is incorporated in the underlying legend.

Keys

Keys to the symbols or patterns used are placed within the map itself (Figs. IV–22, 30). It is not necessary to box the key, although sometimes it is esthetically desirable.

Orientation

By convention, maps are oriented so that the top of the map is approximately north. On maps whose orientation is not readily apparent, an arrow should be included to identify either magnetic or true north.

Insets

Inset maps can be used in two ways. First, on maps involving great detail an inset can locate the main map in a familiar region. For example, a map illustrating a portion of Zaire could have an inset map locating Zaire in Africa. A second use for the inset is to enlarge a portion of the main map in order to show greater detail. For example, on a map of the Caribbean Sea, groups of islands could be enlarged in an inset in order to clarify detail.

Scale

The geographical map should have a bar scale showing distances on the map. Inset maps need their own bar scale. The scale is displayed as a calibrated horizontal line ("bar"), with the unit of distance stated above or below the scale in International System (SI) units abbreviations (for example, in km). The scale is placed in an empty corner of the map, preferably at the lower right. Never use a proportional scale (such as, one inch equals one mile) as these measurements will change if the map is reduced or enlarged.

Figure and ground

To be effective, components of a map must create a visual hierarchy. The most important information must be seen first, and must therefore stand out as a figure in relation to the surrounding background. As an example, islands are generally figures; surrounding seas are background. To achieve visibility, the figure is generally located centrally, and more texture or color is used on the figure. By thus distinguishing between figure and ground, the map becomes easier to interpret (see Figs. IV–24, 28, 29).

Computer mapping

Maps drawn by computer, to be acceptable for publication, must adhere to the same standards as those prepared manually. As each year brings dramatic improvements in computer mapping capabilities, achievement of these standards is becoming more widespread.

Benefits of computer mapping include not only speed and increased efficiency but also the capability to manipulate scale and explore alternative ways of presenting data. Like computer graphing, computer mapping has led to new applications and to the discovery of new relationships. Other benefits include ease of storage and of revision of masses of geographical data.

Limitations to computer mapping revolve around cost and quality. When a great deal of data processing is necessary, or when many maps of one type are needed, the computer can certainly aid the cartographer.

GEOGRAPHIC BASE FILES

To be able to draw a base map on which to display data, it is first necessary to have the geographical information for the base map stored in digital form on magnetic tape. This is the geographic base file. It comprises digital descriptions of geographic areas. The U.S. Bureau of Census, among others, has prepared, and is continually upgrading, comprehensive base files, but these files are computer specific. Translating the U.S. Bureau of Census data files to your computer is at best difficult, at worst impossible.

MAPPING PROGRAMS

Once base map data are stored, it is necessary to code the data to be displayed. Then the computer is instructed in how to manipulate these two sets of data to prepare the desired map. Many programs, also computer specific, have been written for this purpose. Programs exist for preparation of most types of distribution maps—choropleth, point symbol, and isoline, for example.

OUTPUT DEVICES

There are two basic types of graphic output devices, the line plotter and the character or line printer. Maps produced on line printers are often coarse and unattractive, and are not suitable for publication. Maps produced by line plotters are potentially suitable for publication. Users of this book are urged to ask programmers to produce quality printouts suitable for reproduction.

Map Design as It Affects the Printing Process

Line copy

Most maps reproduced in scientific literature are line copy. That is, they consist of black lines or dots on a white background. These are most economical to reproduce. Maps for black and white line reproduction should be supplied to the printer as camera-ready copy. Base map and overlays should be carefully registered, and all instructions should be clear. Chapter III deals with preparation and submission of line copy.

Continuous tone copy

Occasionally a map may be designed as a continuous tone drawing in which black and shades of gray are used. This type of map would be reproduced as a halftone or combination line and halftone figure, and would be more costly. See Chapter III for details of artwork preparation.

Color

There is no denying the effectiveness of color as a tool for increasing contrast and visual impact in a map. However, it has its price. On the original artwork, each color must have a separate overlay. Separate printing plates must be made for each color; special proofing methods must be used. Registration is critical. Chapter III deals with preparation of color copy.

Foldouts and supplements

If the author needs a map of a large area at a large scale, two options are available: foldouts and supplements.

For a foldout, one edge is bound into the backbone of the publication. Thus, the vertical dimension of the map is limited to the height of the page. The horizontal dimension can be four or five page-widths. The resultant format would be useful for a long narrow area, for example, a river.

If desired map size exceeds the measurements of the publication format in both directions, a supplement is the appropriate option. The actual printing and folding costs of a supplement are generally no greater than the equivalent area would be on a per page basis. However, including the supplement with the publication presents difficulties. One solution is to insert the map loosely between the journal pages. Loose inserts, however, are easily lost, so they are best when they serve as a self-contained graphic essay. A better solution for keeping a map supplement attached is the back pocket, but libraries often remove journal covers preparatory to binding. Both methods of inserting supplements are labor intensive and therefore add expense.

In considering any options available for more complex maps, be it color or foldout or supplement, the author must weigh cost against effectiveness. Simplicity is usually the key to effective communication. However, there are some cases in which a complex graphic presentation is the best solution. The expense may not be so great when considered against the many pages of text required to explain the same concept.

Algorithms and Flow Charts

In the computer world, the term "algorithm" refers to a procedure that leads, by a series of choices, to a correct answer. In non-computer applications, the algorithm is a method of breaking down a complicated decision-making sequence into its components. For visual presentation, each step in the series or sequence is presented as a question inside a box (Fig. IV–33). The answer to each question leads via an arrow to a new step. Properly arranged, the sequence of choices and results becomes clear and logical.

Flow charts are diagrams that organize and present processes, sequences, or systems. As in algorithms, each step is written inside a box, and the direction is indicated by lines or arrows connecting the appropriate boxes.

As with all other graphic displays, both algorithms and flow charts can be designed so that they are either quickly understood and easy to use or ineffective and confusing. Both content and design affect final success.

The first steps in preparing an algorithm are writing a clear statement of purpose and preparing a list of objectives. What is the reader supposed to be able to do after using the figure? The next steps are to analyze the task and then to place the constituent parts in boxes and arrange them logically. Simple sentences and active verbs are used.

Constituent parts of algorithms and flow charts are arranged into a logical visual sequence, for example, from top to bottom or from left to right. In an algorithm, each box should ask only one question, and "yes" and "no" should have consistent directions. More important points should look more important. A heavier border or a different shape can be used for emphasis.

After the content and design are finished, the chart is rendered. There are two ways to proceed. Some journals will typeset all the words and have their artists prepare the boxes. Most journals, however, refuse to

FIG. IV–33: *Algorithm for preparing an illustration. (Adapted from Kerlow IV, Rosebush J: Computer graphics for designers and artists. New York, Van Nostrand Reinhold, 1986. Reproduced with permission of the publisher.)*

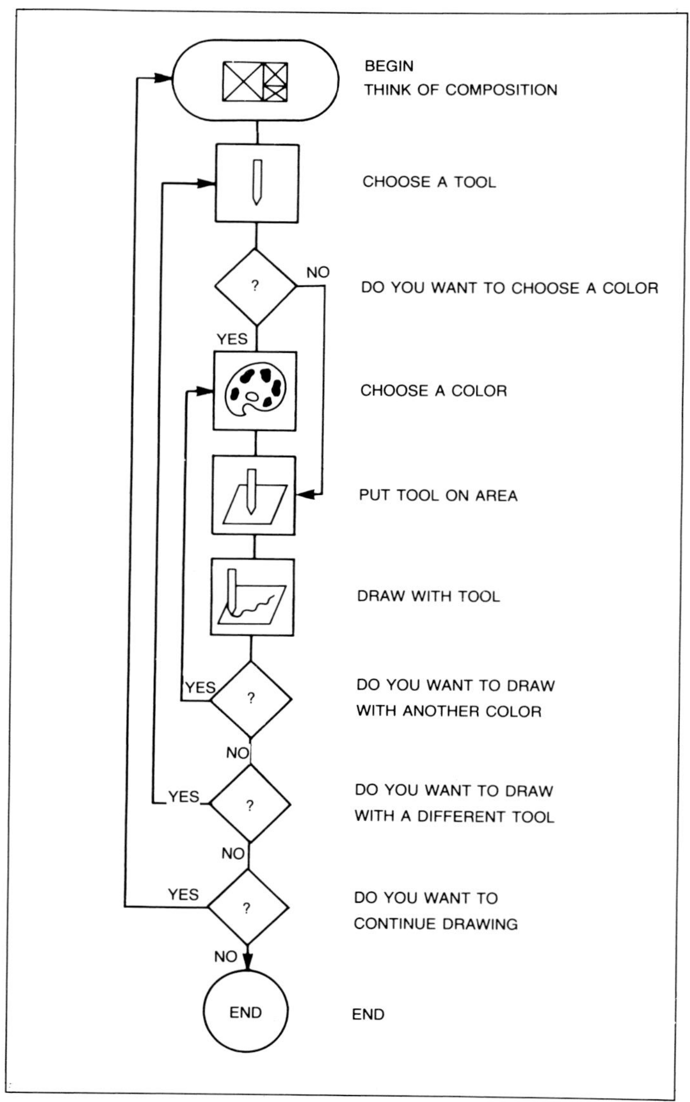

handle this task, considering it the author's responsibility. In such cases, the author should have the entire figure executed by an artist.

To render the chart, determine the column width (or page width, depending on the complexity of the chart). Prepare the original so that it can be reduced to 67% or 50% of the original (see Chapter III). On final reduction, x-height of the smallest lower case letter should be no less than 1.5 mm. Since lower-case letters are easier to read than capital letters, they are preferred.

Legends rather than titles generally accompany algorithms.

CHAPTER V
COMPUTER GRAPHICS

Using Computer Graphics to Illustrate Science

Capabilities and Uses

Modern computer technology provides powerful tools for illustrating the methods and results of scientific investigations. Most graphics that can be produced by hand can be done as well by computer. Although the possibilities are virtually limitless, widespread use of these technologies is currently inhibited by rapid technological advances that render expensive systems quickly obsolete, the high cost of high resolution systems, the substantial learning curve for operating them, and the difficulty of choosing between systems that produce publication quality output and those that do not. Despite these difficulties, computer graphics systems suitable for use by a studio of artists in which expensive systems and training are feasible are commercially available. However, the typical low-cost "business graphics" systems designed for personal microcomputers cannot as yet produce graphics suitable for publication, and systems that encourage the joint development of scientific graphics by scientists and artists sharing graphics files are not widely available. Rapid technological advances, however, promise to make computer graphics increasingly available at lower prices by the early 1990's.

Uses of computer graphics in the biomedical sciences can be roughly classified into three categories: analysis of data collected in scientific studies; display of results in color slides for presentation; and production of publication-quality black and white figures for publication in scientific books and journals. Analysis of quantitative data is the oldest and most highly developed of these uses. With the wide availability of powerful software for analyzing data, such analysis is the mainstay of computer graphics, but the quality of output is rarely adequate for publication. Use of computer graphics for production of color slides is more recent, but because of widespread demand and large economic impetus for business presentations, development of color graphics for this purpose has evolved rapidly. In contrast, the technology for producing publication-quality hardcopy output from computer graphics systems is the least advanced, and remains limited to relatively expensive, high-resolution commercial systems. Although computer graphics techniques for using computer graphics for analyzing data and producing color slides are of great interest to many illustrators, this chapter will concentrate primarily on production of standard black-and-white graphics of publication quality.

Why Use Computer Graphics for Illustration?

There are three main advantages of using computer graphics to produce publication-quality illustrations. First, an illustration, once composed, can be easily modified and printed out again, whereas it is more difficult for artists working without computers to modify a finished illustration by hand. A studio of artists using computer graphics can make corrections and even wholesale changes more quickly and less expensively; this capability encourages researchers to seek perfection in their illustrations rather than settling for the first rendering. Such flexibility is particularly useful for correcting illustrations when last-minute analysis or data editing necessitates changes.

The second major advantage is that, given a suitably "user-friendly" computer graphics software package, the researcher can produce his or her own illustrations without having to interpret the specifications to another person. Although many researchers lack the artistic ability to compose graphics, those who are able can have the process under their own control. This control, together with the ability to make changes rapidly, enables the researcher to evolve a statistical graph through several repetitive versions as the research project progresses and matures, until the final version ideally expresses the scientific point. This process can result in illustrations that display the scientific findings far more powerfully than can traditional illustrations produced in one or two steps.

The third advantage is the ability to display graphics of datasets containing too many data points to be plotted feasibly by hand. This is particularly true of data from experiments producing readings from laboratory equipment that may require the resolution of thousands of points for a short line graph, but it may also be true of clinical and epidemiological data involving large numbers of patients or subjects. In the past, such data were often plotted by the laboratory equipment on crude graphs, or by low-resolution computer output devices, and were then traced into more finished illustrations by an artist. By converting the analog output signals from laboratory instruments to digital data (so-called "A-to-D conversion"), the researcher can use computer graphics to produce the entire graph complete with appropriate axis labeling, titling, etc. With the increased availability of computers and the increasing feasibility of maintaining and analyzing large databases, the need for graphs with large numbers of datapoints can be expected to grow in the future.

The Artist, the Scientist, and Computer Graphics

Although some may view computer graphics as a replacement for the graphics artist, in practice their use has merely enhanced the artist's position as an efficient producer of publication-quality art and as a collaborator in the scientific process. Particularly in cases where computer graphics have been used to produce color slides, studios of artists in academic institutions as well as in commercial firms have generally been the first to acquire high quality computer graphics equipment. Such systems have reduced response time and increased the profitability of producing text slides and artistic graphics. In circumstances where medical artists have mastered the computing process and acquired systems capable of producing publication-quality illustrations, the combination of their artistic skills with computer efficiency has proved a useful mix for most illustrations, including diagrams, standard charts and graphs, and even free-hand drawings that can be input into the computer (Kerlow and Rosebush, 1986). At present, however, most commercial artists' workstations do not easily handle many types of graphs of statistical and laboratory data that must be computed directly from numerical databases.

For production of publication-quality illustrations, artists have for many years collaborated with scientists by using their computer graphs to produce finished illustrations. For example, it has been commonplace for artists to trace lines or other images produced by primitive computer output devices. As laboratory computer output has approached publication quality, it has become more common for artists to accept a computer graphic from an investigator, modify the computer-drawn illustration to overcome limitations in labeling, scaling, etc., and produce a final graph for publication. This collaboration between computer-using artists and scientists can be particularly beneficial because it is often quicker for a scientist to have an artist modify a computer graphic than to figure out how to re-program or "fool" the software to circumvent its limitations.

To encourage this collaboration, some institutions have adopted standard types of paper for computer output that are particularly amenable to alterations by their artists. Alternatively, artists often make photographic reproductions of the raw computer illustrations and make their corrections on the "stats." Although some systems presently allow limited transfer of files by means of floppy disks, the next generation of graphics systems is expected to enable scientists and artists to collaborate more directly by sharing graphics computer files. Scientists will be able to compute the most accurate representation of the analysis and artists will be able to access the same computer file for final artistic embellishment.

Publication-quality Standards for Computer Graphics

Over the past several decades, as computers were increasingly used in scientific research, an infatuation with the relatively complex and expressive graphics that could be produced led many publishers to print computer-generated graphs that did not measure up to the usual quality of manually produced illustrations. Such substandard computer graphics, however, have appeared less and less frequently, until presently most journal editors insist that computer-generated graphics adhere to the same standards as those generated by other means. Therefore, *the standards for publication presented throughout this volume apply equally to computer graphics*.

The main reasons that computer users were led in the past to submit substandard graphics were the frequent limitations of software function and the low resolution of most computer output devices. Software limitations stem from the necessary trade-offs among functionality, ease of use, and cost. Software packages that enable the user to control every attribute of an illustration are generally difficult and time consuming to learn and operate. Conversely, software developers make their software easy to learn and use by building in obligatory default values for many parameters that they believe the majority of users will not want to control. It is precisely these parameters, however, that usually make the difference between "user-quality" and publication-quality graphics. This dilemma is generally solved only by extremely complex and lengthy programming by software developers who have made extraordinary efforts to identify and write specifications for both analytic and publication requirements. This solution is generally very expensive and explains, in part, the seemingly high cost of sophisticated commercial graphics systems.

The resolution of computer output devices presents similar dilemmas (see below). Virtually everyone is familiar with the sawtoothed or "jagged," appearance of diagonal and curved lines on graphics produced by most computers. This lack of smoothness is the main manifestation of low resolution of computer output devices (for example, printers, plotters, film recorders). Precise criteria for adequate resolution have been difficult to set because the only agreed-on standard, often stated in the "Information for Contributors" sections of scientific journals, is that "curves must be smooth." This standard defies quantification because, as will be discussed later, with most output devices (that is, all raster-based devices) smoothness of diagonals and curves can only be approximated; the perfectly smooth curve is not possible. On the other hand, above certain quantifiable levels of resolution, diagonals and curves that appear smooth to the unaided eye can be produced, and even at somewhat lower levels of resolution the "jaggedness" can be overcome by photoreducing the illustration before publication, although there are definite limits to the amount of reduction possible (see Chapter III).

General Strategies for Using Computer Graphics

The most common computer-prepared illustrations for scientific journals and books are statistical graphs and free-hand or diagrammatic drawings. To design and set up a computer graphics system, one must understand the basic strategies used by these systems to produce the finished work.

To create a statistical graph, all computer graphics systems perform a set of basic functions. These include defining the type of graph (for example, line graph, bar chart, pie chart); constructing the axes (for example, scaling, numbering, labeling, defining tick marks); writing and positioning titles, footnotes, legends, and other labels; choosing cross-hatching or coloring of bars or other areas; selecting line types; determining the thickness of axes and lines; choosing and "fitting" regression lines and confidence limits; and other miscellaneous functions (for example, selecting type fonts or plot symbols). Although these functions are usually taken for granted by beginning computer users, a publication-quality illustration usually requires very careful selection of these parameters.

When first using a computer graphics system, the user often wishes to select a few parameters and allow the software's *default parameters* to determine the rest (for example, the scaling of axes, tick marks, line types, etc.). Although graphs can be produced rapidly this way, one cannot expect them to be of publication quality. Acceptable illustrations require the user to become extremely familiar with all of the many parameters and options and to specify virtually all of them in the set-up for the particular graph. The amount of time, effort, and concentration required to master software to this level is usually beyond the time or interest constraints of the scientific investigator. In institutions where particularly "user-friendly" graphics software has been developed for production of basic scientific graphs and charts, however, the necessary expertise can be acquired rapidly, particularly if the software provides hierarchical menus and abundant help screens oriented to the users' level of computer sophistication. Commercial graphics workstations for studio artists can be expected to require around 40 hours of initial training and approximately three months of steady use to develop full proficiency in controlling all of the necessary design parameters.

Besides the strategies for producing the usual statistical graphics, many computer graphics systems allow the development of free-hand drawings, complex diagrams, and a wide variety of artistic effects (for example, shading, "echoing," etc.). Generally, free-hand drawings or tracings are done by means of a digitizing tablet, a flat drawing surface that enables the artist to draw the shape with an electronic pen so that the computer can capture it. Drawings entered into the computer in this way can be saved and recalled for altering, printing, or adding to another illustration. Although a virtually infinite array of artistic effects is possible, these powerful techniques should not be used except when they will substantially augment the clarity, precision, or informativeness of the illustration; they are not meant simply to be eye-catching or dramatic. Overuse of special effects detracts from the scientific purpose of the publication and may well lead to distortion of scientific objectivity.

Elements of Computer Graphics Systems

The elements of a computer graphics system usually include a computer with monitor and keyboard, storage devices, graphics software, graphics input devices, and graphics output devices, each of which will be discussed in detail.

Packaging of the Elements

These elements may be packaged in any of several ways, including a fully integrated graphics workstation with all the elements built into one ergonomic desk unit, distributed systems with many users' display terminals attached to a central graphics-supporting computer by a telecommunications system, or freestanding component parts on the user's desk, connected by cables to the primary system.

In general, the *integrated graphics workstation* is the most appropriate for a graphics studio or department with a single artist dedicated to computer graphics or with several artists who regularly spend time on the system. Its primary advantage is that all elements are conveniently arranged together in a workspace designed for efficient production. Its disadvantages are that it is frequently the most expensive alternative and that technological innovations can be incorporated only when the commercial vendor produces and releases them, often well behind the most advanced technology.

The *distributed graphics system* may be better suited to environments in which large numbers of artists or scientists must use the same graphics systems simultaneously, as in a university or a large research institution. In such a setting, a distributed system avoids the need to purchase many expensive stand-alone workstations and enables a central staff to manage the main computer and software, provide training for a single system, and maintain the most advanced computer graphics technology.

By providing all users with efficient access to the same graphics files, centralized systems are also the most conducive to easy collaboration between artists and scientists for production of the most artistic and scientifically expressive graphics. Their main disadvantage is the need for more powerful central computers capable of supporting many simultaneous users and for central expertise to manage the system. In most universities and corporations, however, these functions are capably performed by organizational computing centers.

Freestanding graphics components, typified by the personal microcomputer, are intended for the individual artist or scientist who is quite knowledgeable about computers and willing to spend considerable time to keep the system current and functional. Their main advantages are that they tend to be the least expensive alternative for an individual and that component parts can be upgraded and replaced as new technology appears on the market, a considerable advantage at the present rate of technological innovation. Their disadvantages are that the required time and computing expertise are presently beyond most artists and scientists and that affordable components often cannot produce publication-quality output.

The Computer

The *computer* is the box of circuit boards, chips, wires, etc. that performs the computing operations. The computer is actually composed of standard parts such as the CPU (central processing unit), a clock which determines how fast the operations are performed, several input and output ports to which the various input and output devices are connected, and a BUS, or broad wire, that connects all of these parts. It has an operating system that manages use of the computer's resources, various programming languages, library subroutines, and program packages to perform graphics and other functions. Familiarity with these terms, although useful, is not absolutely necessary for the average artist or scientist to use most computer graphics systems productively.

There are four basic classes or sizes of computers used for graphics. These are, from the least to the most powerful and expensive: microcomputers (for example, personal computers); minicomputers (for example, the VAX family of computers); mainframe computers (large centralized business computers); and supercomputers (that is, extremely high speed computers with several CPU's running simultaneously in parallel, used primarily in engineering).

Microcomputers, the smallest and slowest but least expensive computers, are best suited for analyzing small sets of data, composing relatively simple graphics, and producing "user-quality," but not publication-quality, illustrations. With late-1980's technology, these small machines are limited by processors that are too slow to run the large, complex programs required for publication-quality output, although by the early 1990's technological innovation is expected to make more powerful generations of microcomputers the most widely used technology for producing publication-quality graphics. Some commercial graphics workstations use microcomputers in their turnkey systems; although they may function satisfactorily for their intended applications, they generally support only one user at a time per machine and have generally primitive means for artists and scientists to share graphics files. Trying to support more simultaneous users or to run adequate software often yields prohibitively slow performance.

For wider graphics applications or for supporting many simultaneous graphics users, *minicomputers* are presently the machines of choice. These computers are fast enough to run very sophisticated graphics software, and many of the most powerful software packages have been developed for minicomputers. At present, the best commercial multi-user graphics systems are based on minicomputers, as are most systems developed by universities and corporations for internal use.

Although much early graphics software was developed for *mainframe computers*, these systems have not kept abreast of the trend towards user-friendliness for artists and scientists. With a continuing heavy orientation towards business applications, mainframe software remains difficult to use, requiring experienced programmers to develop graphs, and is not particularly suited to publication-quality graphics. In addition, with the intense competition from micro- and minicomputer developers, it appears unlikely that mainframe computers will develop rapidly in this direction.

Supercomputers, which perform the most sophisticated and complex graphics functions possible, are primarily used for computer-aided design (CAD) in engineering and related fields. These ultrahigh-speed machines compute two- and three-dimensional graphics images too complex to be generated on mini- and mainframe computers, and make possible simulation and testing of engineering designs entirely on computer. Although images generated on supercomputers are sometimes used as illustrations for scientific articles, their extremely high cost prohibits their use for routine illustration.

The Monitor and Keyboard

The user generally gives the computer instructions and views the initial results of the computing operations on a computer screen (similar to a television screen), usually called a *monitor* or *CRT* (cathode ray tube), to which is attached a *keyboard* similar to a typewriter keyboard. Special devices, such as the "*mouse*" and *thumbwheels*, supplement the keyboard by increasing the speed with which commands are given or the precision with which the screen can be modified. Technically, the monitor with keyboard is both an input device (for entering data and instructions) and an output device (for displaying the results). It is considered separately here because it plays such a central role in computer use and because its output is rarely used for publication purposes.

With most graphics systems the user chooses the parameters for the graphic by entering commands on the screen or pressing certain *function keys* (usually labeled "F1" through "F16") on the computer keyboard. Once the graphic is constructed by the computer, using the parameters selected, it is displayed on the monitor to show the user the result of the parameters chosen. In most cases, the image is not satisfactory on the first try; additional parameters are then chosen and the image is again displayed on the monitor, and so on until the graphic is satisfactory. In this way the monitor is used for repetitive input and output to compose a finished graphic, which is then sent to a high-resolution output device (see below) for production of the final publication-quality illustration.

It must be emphasized that the monitor is virtually never used as a final output device for publication-quality graphics. Although the technology for photographing the monitor is available, it is almost never used because most monitors do not have sufficiently high resolution and are subject to serious distortion caused by the convex curvature of the screen and interference from room light. That monitors are not used for final output does not, however, obviate the need for reasonably high resolution. Such resolution is needed in composing graphics to ensure accuracy of detail—for example, to join two parallel line segments so that they meet exactly, or to modify free-hand sketches precisely. Therefore, even though inexpensive low-resolution monitors are widely used for business graphics on personal microcomputers they are generally not recommended for publication-quality graphics.

A recent important innovation is the development of *monitor emulation packages*, software packages that can enable an inexpensive monitor to perform many of the functions of a far more expensive graphics workstation. Of particular interest are the graphics emulation packages written for personal microcomputers with enhanced graphics monitors in multiuser graphics systems. With these packages, most users can compose their graphs on low-cost microcomputers and merely check the final product on a single, centrally located high-resolution monitor. Although lower-resolution screens of microcomputers in emulation mode may cause occasional misalignments and esthetic problems in illustrations, they are often quite attractive from the standpoint of cost effectiveness.

Storage Devices

All computers must have attached *storage devices* with which to store data and programs. Programs include the software systems needed to operate the computer (systems software, or operating system), the applications programs (for example, computer graphics packages), and the specific graphics files (lists of parameters that cause the graphics software to draw the graph). Each file containing data or programs is stored under a standard *file name*, a descriptive name usually written in all capital letters and followed by a period and a three-character *extension* that classifies the file (for example, GRAPH3.GRF). All files are listed in a *file directory* so that they can be easily located and recalled.

Four basic types of storage devices are in common use: floppy diskettes, internal hard disks, attached disks, and system disks. *Floppy diskettes* and internal *hard disks* are almost always used with inexpensive microcomputers. Floppy diskettes are often used when a scientist wishes to transport a file of data or a graphics file to an artist's microcomputer for artistic refinement. Attached disks can be added to a small computer to increase substantially the amount of data or the number of programs stored. Centralized mini- or mainframe computers almost always have large volume *attached* or *system disks* to store very large amounts of data and software. The type and size of storage devices required depend on the amount of information to be stored and the type of computer being used.

All storage devices suffer from one serious fault: if they become damaged they may cause all the stored information to be lost. For this reason the contents of all disk storage devices must be periodically copied to a backup disk for safekeeping, to ensure against loss of the information.

Graphics Software

The term *software* denotes the computer programs that enable the user to make the computer hardware execute the desired operations. A piece of software generally consists of a long text file of written *commands* in a programming language (for example, BASIC, Assembler, Cobol, Fortran). Stored on a disk, software is *loaded into memory*, or *initialized*, either when the user turns on the computer or when a loading command is entered. During initialization, the many commands in the software file are read from the disk and stored in the computer's active memory, where they remain ready for use until unloaded from memory or until the computer is turned off. As long as the software is loaded in memory, the user can enter commands and the software will execute them by sending the appropriate impulses that drive the hardware.

Computer graphics software is written by programmers, or companies of programmers, who try to build in all the functions they expect the user to require. These include the ability to input data, construct different types of charts or graphs, control the width of lines, choose the desired fonts, send the image to an output device, etc. When buying software, however, one must be aware that unless the software is specifically designed to perform a given function, it probably will not be able to do so. Consequently, for computer graphics software to be functionally adequate it must be designed by highly skilled developers who are quite familiar with the needs of the target user group. Generally, a software package specifically directed towards a given user group will be more functional and easy to master; conversely, a package designed for a diverse user group is likely to be less functional and more difficult to use. The functions available and the instructions for use of a software package are explained in a manual, usually referred to as the *documentation*. Good documentation is clear, concise, complete, and written in language that can be understood by the target user group. Poor documentation often exposes poor software.

There are three basic levels of computer graphics software. These are programmers' basic "tool kit" software, user interfaces, and peripheral device drivers. Computer graphics *"tool kit" software* includes very basic programs that directly control the computer hardware functions. They tend to be functionally very general and rather difficult to operate, usually requiring a highly experienced programmer to produce finished graphics. However, these software packages are not ordinarily used by artists and scientists. Instead, other programmers, often those in the organization using the package, develop a *user interface* software package that runs along with the "tool kit" package to make it easier for those with less computer training to use. Whereas the usual basic "tool kit" package requires programmers to type in complicated commands to run it (a *command-driven system*), a typical user interface provides menus for the users to select the desired parameters (a *menu-driven system*). Command-driven systems are generally far more versatile and powerful but difficult to learn, whereas menu-driven systems tend to restrict the functions to those needed by the user groups but are far easier to learn and use.

Peripheral *device drivers* are software packages that enable the user to operate peripheral devices, such as printers, plotters, digital film recorders, and special graphics monitors. They customize the graphics image produced by the main graphics software into an image that can be displayed by the particular peripheral device. Monitor emulation packages represent a special type of device driver. Device drivers are often built into software packages but may also be marketed separately. When planning a new computer graphics system one must be certain that the appropriate device drivers are available to run the desired peripheral devices, which will not run satisfactorily without the right driver.

The sophistication and adequacy of computer graphics software packages vary widely; most software packages are not capable of producing publication-quality illustrations. In general, a package's level of sophistication is related to its size and cost; truly functional packages tend to be very large, occupying large amounts of space on disk storage units, and tend to be expensive. Conversely, inexpensive packages that fit on one or two floppy diskettes are generally not adequate for producing publication-quality illustrations.

When evaluating computer graphics software, the prospective buyer must be extremely cautious and thorough to obtain a system that will suit the particular environment and produce the desired results. In general, the factors to assess are graphics functionality, user efficiency, and machine efficiency. *Functionality* includes all of the graphics functions that will be needed. The buyer must be assured that the package can generate all or most of the graphics required, that all of the parameters necessary for high-quality illustrations are under the user's control, and that the system will achieve the level of resolution required for publication.

Assuming adequate functionality, the next most important quality is *user efficiency*, the ease and speed with which an average user can operate the primary features of the software. Unfortunately, wide functionality and user efficiency tend to be competing priorities for software developers, and thus few packages possess both qualities. A highly functional system is of little use, however, if it is too difficult to operate or if user training is too time consuming and costly. The following features of the user interface tend to increase user efficiency. Menu-driven interfaces are generally far more efficient to use than command-driven ones, particularly if the menus follow a logical "tree" structure that can be self-teaching. Default values for many of the parameters can be set to artistic standards to reduce rote input. Menus or value tables allow summary views of the parameters selected to increase the efficiency of editing and altering a graphic. A graphic can be displayed rapidly at virtually any stage in its development to show the user what the current parameter selection has thus far created. Menu selections or commands are precise enough to minimize the need for trial-and-error composition. "Mouse" or thumbwheel controls can be used for manual sizing and positioning of titles, labels, special characters, legends, etc. on the screen. Several graphics can be easily combined to produce a complex graphic. Collaborating artists and scientists can easily share a graphics file under development. Graphics can be generated not only from data typed in by the graphics user but also directly from large data files stored in the computer. Safeguards are provided to prevent the user from inadvertently deleting a graphics file and to allow rapid saving and backup of files. Documentation is sufficient for the user to locate any needed function rapidly and to master it without the need for consultation with the software company.

Finally, good graphics software must have a high degree of *machine efficiency*, the ability to perform the needed functions without overtaxing the hardware for which it was designed. In the past, software developers tended to value machine efficiency far above user efficiency, which explains in part the difficulty of using many of the older command-driven systems. With the greater competition among software products and the great increases in computing speed and memory capacity, however, user efficiency has rightly taken a higher priority. Still, before investing in an expensive graphics system, one should be satisfied that the software will support the required number of concurrent graphics users, doing a representative mix of functions, on the computer hardware suggested by the software developer.

Special Graphics Input Devices

By far the most common method of entering ("inputting") information into a computer is by typing characters or pressing function keys on the computer keyboard, to either enter data or to give instructions to the software resident in the computer's memory. In addition to the keyboard, special input devices can be very useful in developing computer graphics.

The special input devices most often used are the digitizing tablet and the image scanner. A *digitizing tablet* is used to enter a free-hand drawing or a tracing into a computer file, where it is automatically transformed into a computer-readable data file and stored on the computer disk. An *image scanner* is a piece of photographic equipment that takes a photograph of virtually any object, picture, or other image and transforms the image into a computer-readable data file. Image files entered into the computer by either of these special input devices can later be recalled from the storage disk to be modified, artistically embellished, or combined with other images, graphs, or text.

Graphics Output Devices

A graphics *output device* is any of a wide variety of computerized devices that produce the actual visual images which constitute the final product of computer graphics composition. Examples include CRT screens, pen plotters, dot-matrix printers, ink-jet printers, laser printers, digital film recorders (electronic cameras), thermal printers, and computerized typesetters. All graphics output devices are operated by software packages, called device drivers, specifically designed for the particular device.

All commonly used output devices operate on one of the two following basic principles: raster (dot) processing and vector (stroke) processing. Basically, pen plotters operate on the vector principle, and virtually all others use the raster principle in one form or another. *Raster processing* originated with early television technology in which a ray of electrons is scanned along horizontal lines illuminating tiny phosphorescent dots, or pixels, on the screen. Today the term is applied to any process in which an image is formed with tiny dots, whether the dots are produced by a raster scanning gun or some other method. *CRT screens* use raster guns to light up phosphorescent pixels. *Digital film recorders* provide a 35 mm camera carefully matched to a special high-resolution CRT screen to produce high quality photographs. *Dot-matrix printers* use a print head with a rectangular matrix of print wires that strike the paper through an inked ribbon. *Thermal printers* are similar to dot-matrix printers except that the print wires are much finer and use heat to imprint on the paper. *Ink-jet*, *wax-jet*, and *plastic ink-jet* printers spray ink, wax, or plastic pigments in tiny dots on the paper. *Laser printers* use a laser beam to expose dot patterns on a photosensitive surface that attracts ink powder and binds it to the paper. The key point is that all of these raster devices produce patterns of dots that together make a graphic image.

In contrast, *vector devices*, that is, pen plotters, mechanically draw lines on a paper surface with ink pens. The pen is driven by mathematical algorithms in the graphics software (usually the device driver). The algorithms view the plotting surface as a rectangular area mapped off in Cartesian coordinates (a horizontal and a vertical axis) with an infinite number of points on each axis. (In contrast, raster areas are mapped into a discrete number of points, or pixels.) The plotting algorithm starts the pen at an exact point indicated by a pair of X and Y coordinates and directs it to draw a line of specified length and curvature. Sophisticated plotting algorithms can draw virtually any shape with great precision. The final plots can then be photographed to produce glossy prints for publication.

In addition to the functionality of graphics software, the most important determinant of the ability to produce publication-quality computer graphics is the *resolution* of the output device. The resolution of vector devices is determined by the level of mechanical precision (usually very high, for example, 0.0025 mm) and, more important, by the sophistication of the mathematical algorithms in the graphics software (usually the device driver for a pen plotter). In contrast, the resolution of raster devices is determined most importantly by the density of the dots, or pixels and, to a lesser though sometimes important extent, by the sizes and shapes of the dots.

Although no standards for resolution levels of raster devices for publication quality have been formally established, approximate ranges of acceptability can be suggested. For CRT screens and digital film recorders, both of which form their images from lighted pixels on a CRT screen, publication-quality images usually require the level of resolution produced by a screen with over 4,000 pixels on each horizontal scan line and at least 2,000 scan lines. Dot-matrix, laser, ink-jet, and thermal printers that form dots on a page should generally produce more than 500 dots per inch (dpi).

With the technology of the late 1980's, the most satisfactory output devices for producing publication-quality illustrations are vector pen plotters and high-resolution digital film recorders. For line drawings and most statistical graphs, pen plotters are the least expensive and most widely used devices. Recent improvements in quality and sharp reductions in price of digital film recorders have made them the most attractive alternative. At present, the widely available dot-matrix, laser, ink-jet, and thermal printers produce either poor resolution (150 dpi) or marginal resolution (300 dpi) images. At the opposite extreme, thermal dot-matrix typesetting devices produce traditional, ultra-high resolution (1,200 to 2,400 dpi). New generations of printers that produce moderate though acceptable resolution (600 dpi) are expected by the end of the decade, although prices may initially keep them out of financial reach for many scientific art environments (Stewart and Tazelaar, 1987).

Color Graphics

The use of color graphics to date has been largely restricted to production of color slides for presentation. Color graphics have not been developed extensively for use in publications, for several reasons. First, the widespread practice of photocopying journal articles and the limited availability of color photocopy machines reduces the usefulness of color illustrations unless they are also suitable for black-and-white copying (for example, color-coded lines also using different line types). Such adaptations are usually considered unesthetic. Second, color illustrations are more expensive to print, and the extra charges are usually passed on to the author, thus producing a financial disincentive. Third, until recently output devices for producing publication-quality color graphics have been of limited sophistication. As a result, whereas color photographs (for example, color photomicrographs) have appeared increasingly in publications, color graphics are still relatively uncommon.

Three basic types of color output devices are capable of producing publication-quality color illustrations. These are pen plotters, ink-jet (or wax-jet) printers, and photographic prints from color slides (or the slides themselves when accepted by the publisher). Pen plotters are mechanical raster devices employing from 4 to 16 (or more) ink pens of different colors, each of which can be selected in the process of automatically drawing a graph. Most plotters in current use are flatbed plotters, which require the user to mount a piece of drawing paper on a flat surface each time a graph is to be drawn. More recently, automated plotters with continuous paper feed allow multiple graphs to be queued up to the printer to be drawn in sequence. The main advantage of the pen plotter is that relatively fine resolution can be obtained. Its drawbacks are that the colors are limited to those of the pens available from the manufacturers, and solidly filled spaces on the graphs (for example, bars, backgrounds, pie wedges, etc.) are often difficult to make acceptably homogeneous.

Ink-jet printers involve an output device that sprays red, green, or blue ink from small jets onto the plotting paper. The color sprays are mixed to produce shades of color determined by the software applications. Problems of this relatively new technology have included messiness of ink handling, smudges on the output plotting paper, and clogging of ink jets with dried ink. A recent innovation has been the development of wax-jet printers that actually lay on the paper a thin coat of colored wax. Although rather new, this technology appears to provide much finer resolution and largely overcomes many problems of the older ink-jet printers.

Making color photographic prints from color slides (positives) or color negatives is the most effective method of producing publication-quality color illustrations. Although either slides or color negatives can be used, color slides produce generally better photographic color prints. The major drawback of producing color prints is related to the technology of making color slides from computer images. The color slides can be produced by simply photographing the computer screen with a 35 mm camera. This is best done in a completely dark room with the camera on a tripod placed at least 10 feet from the screen and using a zoom lens. The drawback is that the curvature of the computer screen unavoidably produces a curved distortion in both planes, such that vertical and horizontal lines appear to be slightly convex to the center. Some computer graphics workstations use mechanical adjustments to compensate for this distortion, but even so this technique rarely provides acceptable publication quality.

Alternatively, color slides can be produced by a digital film recorder, basically a computerized camera system. The computer graphics system produces a graphics output file which is transferred to the digitizing camera, which in turn produces the graphics image in its own internal computer screen which exposes the film three times with a red, green, and blue filter, respectively. The film recorders are engineered to avoid the distortion of standard computer screens. There are two major drawbacks to digital film recorders. First, a particular problem with lower-priced models, film recorders tend to produce "bleeding" of reds and to some extent of blues, that is, colors that contain red (for example, orange) may have red halos around them caused by light scatter. This problem is largely overcome by the much higher-priced, high-resolution film recorders. Second, the colors obtained in slides and prints from digital recorders are generally not true to the colors seen on the graphics computer screen, and these differences can be troubling. This can be compensated for by keeping a light box, or sleeve, of color slides showing examples of slide colors produced by specific numerical settings of the color specifications in the software. Alternatively, one must work out the colors through trial and error; however, given the usual hours or days required to develop color slides or prints, it can be quite time consuming and costly to develop an esthetic color selection this way.

Summary

Computer graphics presently offer an exciting array of powerful tools for creating high-quality scientific illustrations faster and more efficiently than in the past. Although seemingly complicated, computer graphics systems are composed of a definable set of elements that are connected in standard ways and operate up to reasonably measurable specifications. When setting up a successful computer graphics system, one must understand these basic concepts and thoroughly study prospective systems to ensure that they will fill the needs of funtionality, user efficiency, and machine efficiency. However, since the market is flooded with computer graphics systems that cannot produce illustrations of publication quality, and since high-quality systems are both uncommon and often expensive, one must carefully assess the quality of the illustrations produced and fit the type of system to the specific needs of the organization. Continuing technological advances promise to make even more efficient computer graphics systems available at lower cost in the near future.

Organizations

In the rapidly progressing field of computer graphics, current information is most likely to be found through monthly publications, annual conferences, and regional and local interest groups. The following national organizations hold annual meetings and publish relevant periodicals.

Special Interest Group on Computer Graphics (SIGGRAPH)
Association for Computing Machinery, Inc.
1133 Avenue of the Americas
New York, NY 10036
(212/265-6300)

IEEE Computer Graphics and Applications
IEEE Computer Society
10662 Los Vaqueros Circle
Los Alamitos, CA 90720
(814/821-8380)

National Computer Graphics Association (NCGA)
2722 Merilee Drive, Suite 2000
Fairfax, VA 22031
(703/698-2000)

LITERATURE CITED

Kerlow IV, Rosebush J: Computer Graphics for Designers and Illustrators. New York, Van Nostrand Reinhold, 1986

Stewart GA, Tazelaar JM (eds.): Printer Technologies. Byte, 12 (no. 10):161-230, 1987

CHAPTER VI
CAMERA-READY COPY

Fast, economical publication is vitally important to authors, editors, publishers, and printers alike, and this chapter will consider two methods for achieving these objectives. Both are ways to prepare *camera-ready* text and illustrations complete enough for the printer to make plates to go directly on the press. Camera-ready text must not be confused with *camera copy*, which is copy photographed as part of the prepresswork process in the printshop. Desktop publishing is comparatively new, whereas camera-ready text and illustrations have been in existence for many years.

Desktop Publishing System

The word "publishing" is misleading as a description of this system because desktop publishing really refers to the use of a computer that has three basic programs: word processing, graphics, and page make-up. The final camera-ready copy is produced with a laser printer whose product is nearly of typeset quality. Laser-printed copy can therefore be used as camera-ready copy. It is no longer necessary to go to the drawing board and to cut and paste according to the specifications provided by the printer or publisher as one did when preparing camera-ready copy on an electric typewriter or computerized word processor.

Advantages

The National Composition Association (NCA, a division of the Printing Industries of America) analyzed the division of work involved in phototypesetting including keyboarding, proofreading, corrections, make-up, typesetting, and proofing. Keyboarding accounted for 30% of the labor. Keyboarding, proofreading, and corrections comprised 66% of the work. These three tasks are ideally suited to performance on a word processor before make-up and typesetting. Preparation of camera-ready material with a desktop publishing system can result in significant savings.

The desktop computerized system has a software program to produce simple graphics. The type fonts available from the system are similar to those used by many scientific publications: Times Roman, Helvetica, Narrow Helvetica, New Century Schoolbook, Palatino, and Symbol.

Desktop publishing technology excels at manipulation and creation of graphic elements. Complex charts, graphs, and drawings usually require conversion to a digitized image with a scanner, but once available in a digitized memory, the graphic image can be used, enhanced, and merged with the text electronically.

Desktop publishing allows quick editing, rapid changes, and quick turnaround. Using this system one can produce on a desktop printer approximately a dozen pages with text and illustrations in less than 30 minutes. This is possible because of the memory capability of the program. The system also puts aesthetic and page make-up control in the hands of the user.

Software is the key to satisfying desktop publishing needs. Once the degree of involvement is selected, hardware options can be purchased, leased, or used on a time-sharing basis from a growing number of desktop publishing services. Floppy disks and illustrations from any system can be provided to these service people who can produce camera-ready copy in page form overnight.

Disadvantages

Unless camera-ready copy for desktop publishing is compatible with the publisher's or editor's specifications and the printer's technology, it may be more expedient and economical to use traditional typesetting methods. Furthermore, editors and authors who wish to use the desktop computer to prepare copy for submission to the publisher must not only buy equipment and add to it as necessary, but also must train their typists to produce camera-ready material.

The quality of copy produced by desktop publishing systems is limited by the quality of the printer incorporated into the system. Text produced by high-quality dot-matrix printers may be acceptable for camera-ready copy, but graphics produced by such printers are rarely so. Digitized images produced by laser printers with at least 300 dot-per-inch output *may* be acceptable, but the standards required by editors, publishers, and printers must be met.

Because of the continuous development and recent nature of the market, there are few sources of information on desktop publishing other than the brochures and personal computer publications of the manufacturers (but see Literature Cited).

Camera-ready Text and Illustrations

Camera-ready material is ready to be photographed for plate making. Two reasons for using camera-ready material are to save time and to cut publication costs. In addition, special circumstances, such as complex formulas, tables, and figures, often make the use of camera-ready copy more attractive than having the printer set type or generate illustrations, even if no economies of time or money are involved. In making the decision to provide camera-ready copy, the editor, author, and publisher must balance the advantages of speed, cost savings, and other desirable goals against the disadvantages of a less aesthetic appearance of the final product and the much greater amount of care required in preparing letter-perfect copy for submission to the printer. In the ideal situation, the printer can photograph camera-ready pages with no special preparation. Any portion of the copy that requires separate handling raises the possibility of error and also reduces the benefits provided by this method.

Material Suitable for Camera-ready Text

Symposium proceedings, agendas and abstracts for meetings, "rapid publication" journals, books, and newsletters all benefit from the most speedy publication possible. In addition, by using camera-ready material in reduced (miniprint) size, it is often feasible to include in a primary journal publication some material that would otherwise go into a "depository," where it is stored and available to interested readers only after the considerable delay (and expense) involved in writing to the depository and waiting for the requested data.

Rapid communication

Many publications reserve a "rapid communication" section for articles that are processed as camera-ready text and scheduled separately from the typeset portion of the publication. These articles usually represent preliminary reports or studies that may be of immediate benefit to the scientific community and should be published with as little delay as possible. Such works, therefore, bypass the usual routine of typesetting and paging by the printer and are submitted by the author in camera-ready form.

Miniprint

Miniprint is an effective way of presenting supplemental material, such as extensive data, detailed methods, derivations, or discussions, which can be reduced to a very small size. It is also useful for publication of a large number of abstracts that must be compressed into a minimum number of pages. This type of camera-ready copy can be reduced by 40% to 50% of the original size: however, at 40% the use of a hand lens is recommended for reading (See Fig. VI-1).

The guidelines that follow apply to any material submitted to the printer for camera-ready copy. However, forms and instructions must first be obtained from the editorial office of the specific publication for which the work is intended. These forms are preprinted, and the instructions include specifications for widths of columns, headings, and paging (See Instructions and Forms, page VI–00.)

Judging Original Material for Camera-ready Suitability

Straight prose text is best suited for this process. Transactions and proceedings of meetings and the main text of scientific articles qualify in this category, as do meeting abstracts prepared for later presentation. Tables of data, line drawings, graphs and, in general, any linear representations can also reproduce well. Other candidates are directories, newsletters, indexes, programs, and abstracts.

Considerable care is necessary in preparing complex reaction schemes, chemical diagrams, and mathematical formulas. These are often hard to make uniform in style or blackness. If additional special characters, such as Greek letters, many levels of sub- or superscripts, or chemical symbols, are required, even greater care is needed.

Material Not Suitable for Camera-ready Use

Among the cases that are inherently unsuitable for camera-ready use are high-resolution electron micrographs, color illustrations or any continuous tone photograph, or artwork that requires a multiple-step printing process. Another category of unsuitable material includes anything that requires many changes at the last minute or in which different type styles are desired for aesthetic reasons.

Speedy Publication

Because the material presented to the printer has been proofread in advance by author or editor, not only is the time required for typesetting eliminated but also the time needed for proofreading, except in a general way to check the quality and order of presentation of large pieces of material. The camera-ready material can go directly to the camera department, thus expediting the printing schedule, unless the printer is being asked to assemble different kinds of material (text, formulas, headings, and so forth) at the last stage.

Cost Savings

In the best case, a page of camera-ready copy costs no more than a black-and-white figure to produce because it is laid in front of a camera and photographed in much the same way. There is no expenditure for proofreading, page make-up, alterations, and little cost, if any, for mailing proof to authors. To realize these savings, authors must be required to provide camera-ready material in final form.

The economies are sacrificed if the printer must make any kind of change after the camera-ready copy is submitted. For example, even if the printer must add only page numbers and running heads to the supplied copy before it is shot, there will be a charge, and if the total page must be assembled, the cost will continue to grow. Should the author decide to make alterations in the copy, the corrected sections must be rephotographed and inserted into the prepared material, all of which will add significantly to the cost. A page of camera-ready material would normally cost less than half the cost of a page that must be typeset, made up, and proofread, plus postage for mailing proof to the author and editor. If extensive changes are made to the camera-ready copy, the savings might very well be wiped out.

If editors do not require authors to provide the final form of the camera-ready material, "hidden costs" will be generated within the editorial office if much copyediting and retyping is required, or if preparation of the copy on the office printer is needed to produce uniform final copy. Editors must weigh the aesthetic advantages to be gained against the time and the cost disadvantages arising from the editorial office.

Economy of space

Camera-ready copy that has been typed single-spaced and reduced to 60% of original size will contain about 3,000 words on the same page that would accommodate only about 1,000 words if set in 7-point conventional type. The same material reduced to 40% of original size allows about 4,500 words to fit in the same space. Needless to say, a certain minimum amount of material should be included in each section if it is to be cost effective. However, a very large table, for example, if it can be reduced to fit on one page instead of spreading over two pages, might justify the use of reduced camera-ready copy for that table only. Likewise, by using reduced size camera-ready material, large amounts of data that might otherwise be relegated to a data repository can be printed in the primary publication, with relatively little expense and with the enormous advantage to the reader of having it readily available for reference.

Special Circumstances

Complex material

One significant advantage to authors and editors is the elimination of proofreading. Not only may this speed publication, but in the case of material containing very complex mathematical equations, tables or figures showing amino acid or nucleic acid sequences, or anything in which errors in transcription are easily introduced and difficult to detect, there is an intrinsic value in having such material printed without having to be verified in detail once it has been carefully prepared by the author.

Circumstances do arise where copyediting and typesetting a large amount of material from different sources (the abstracts for a large meeting, for example) are just not practical, even if cost is not a factor. The editorial office can supply preprinted sheets for uniform presentation, and guidelines for copy editing if desired.

The necessity of sending proofs to a large number of authors can be avoided if the authors are asked to supply camera-ready copy.

Visualization of final product

Many editors and authors may find it desirable to be able to visualize the final product, in terms of page layout, from the start. By providing camera-ready pages, they can control the placement of figures, tables, or equations without costly rearrangement of page proof.

Preparation and Handling of Camera-ready Copy

Once the editor or author decides to use the camera-ready technique of publishing, it follows that all the parameters selected should be directed toward maximum economy, facilitating the copy preparation, and expediting the production and printing. The primary advantages of using camera-ready copy are simplification and acceleration of the printing process; therefore, the camera copy should be complete and in optimal condition for the printer. When the process is hindered by the need for extra handling, such as editing on pages or a complex mix of typeset and typescript or illustration inserts, the advantages of using this technique are diminished.

Photographic quality of camera material

All materials to be photographed should be letter perfect; the typescript density should be uniform and black; other ribbon colors, such as brown or blue, should never be used. The paper used for page flats, typescript, illustrations, or other inserts should be selected in advance to be as close to the same shade of neutral white as possible. Any undue differences

FIG. VI-1: *Examples of various camera reductions. A. Original size (100%). B. 80% of original size. C. 60% of original size. D. 40% of original size. D should be used for special purpose miniprint publications only.*

The pages for the miniprint section are supplied by the author in a camera-ready form, with
figures and tables inserted in the appropriate
places in the text. Each legend should follow
its figure (do not group legends together at the
end). Figures should be numbered in the order
cited, whether in the typeset part of the paper
or the miniprint; the same numbers should not be

100%

The pages for the miniprint section are supplied by the author in a camera-ready form, with
figures and tables inserted in the appropriate
places in the text. Each legend should follow
its figure (do not group legends together at the
end). Figures should be numbered in the order
cited, whether in the typeset part of the paper
or the miniprint; the same numbers should not be

80%

The pages for the miniprint section are supplied by the author in a camera-ready form, with
figures and tables inserted in the appropriate
places in the text. Each legend should follow
its figure (do not group legends together at the
end). Figures should be numbered in the order
cited, whether in the typeset part of the paper
or the miniprint; the same numbers should not be

60%

The pages for the miniprint section are supplied by the author in a camera-ready form, with
figures and tables inserted in the appropriate
places in the text. Each legend should follow
its figure (do not group legends together at the
end). Figures should be numbered in the order
cited, whether in the typeset part of the paper
or the miniprint; the same numbers should not be

40%

in image density or any mix of off-color white paper will affect camera exposure; such copy will reproduce unevenly and probably result in obscure or illegible text. Transparent tape, correction fluid, eraser smudges, correction cut lines, and uneven rules will all appear as blemishes when the copy is photographed.

Format selection

Trim size is an editorial determination usually based on the publisher's line of books or journals, and has no inherent effect on camera-ready copy other than the limits set for width and depth of the type page.

Whether to make the text single column or double column is determined by the space available for the type page. The single column page offers advantages of directness which simplifies the procedure of paging text material. When a word processor is used, it is possible to program the output directly into pages. Even if the material contains a few illustrations or table inserts, the necessary space can be allowed for in the text stream as the printout is made.

If page width allows, the text should be set as double column material. This arrangement provides the most efficient use of space and presents more words in fewer pages. However, the double column page must be produced by the cut-and-paste method, usually from random, galley-like proof. If this material is produced by the word processor and has few or no inserts, it is possible to produce the text in exact column depths to facilitate cutting and pasting text material for page make-up.

Type size selection

Legibility studies have shown that typescript size and number of characters per line have a definite effect on the readability of the text. The rule of thumb to follow is to use about one and a half alphabets per line, i.e., about 39 characters. This number is intended only as a guide; some slight variation, of course, will not cause poor readability. The result is that double columns (narrow-width text) should be planned for smaller typescript and single columns (wide text) for larger typescript.

Text material

Copy produced directly from a typewriter is referred to as ribbon copy. To achieve crisp, uniform copy, a "single-strike" high-density carbon ribbon designed for one-time use is recommended. This cartridge ribbon is available for both the typewriter and the word-processing printer. Multi-strike ribbons or cloth ribbons designed for repeated use should be avoided. Uneven density produced by poor ribbon quality will appear as bold lines and light lines with broken letters.

Text for camera-ready copy can be produced by either a typewriter or a word processor. Because the word processor provides so many advantages, it should be used if possible. Once the initial input is in the word-processor memory, it can be edited and reworked as necessary, and preliminary printouts can be used for editing and proofreading. Finally, the text output can be programmed and original copy produced by using the word-processor printer and going directly into completed pages or exact column lengths for two-column make-up.

Not all typescript faces are suitable for reproduction. Some designs are too condensed and reproduce poorly in the printing process. In general, the sizes of typescript available on the typewriter or word-processor printer are larger than typefaces normally used for typeset text. Legibility of typescript is generally improved by reducing the camera copy 10% to 25% of original size. Therefore, the pages are made oversized, and the printer then makes the reduction as necessary in the printing process. Typewriter faces are either monospaced or proportional spaced. The proportional faces resemble typeset faces in design and will produce the most legible text for camera-ready copy.

Special characters, such as Greek letters, accents, or symbols not on the keyboard, must be inserted on the ribbon copy by hand. An extensive library of special characters is available on sheets of transfer type (see Chapter III).

Illustrations, tables, and other inserts must be fitted to column widths and inserted in the oversized page. This will avoid extra camera work and stripping by the printer.

Paper that is smooth and neutral white provides an excellent surface to produce camera copy. An example is Jamestown Xerographic copy paper, which is an inexpensive writing grade bond. This is a "two-sided" sheet, that is, one side is slightly smoother than the other; the smooth side should be used.

Rag bond, crackle or linen finished bonds should never be used, because the textured surface will cause uneven and broken letters in the camera copy. Coated or glossy paper should also be avoided.

Illustrations, tables, and complex schemes or formulas

Illustrations, whether line drawings or halftones, and most tables are processed as separate units and handled as individual inserts. To facilitate printing production, all inserts are sized and fitted into the page flat along with the text material.

Line drawings, diagrams, charts, and graphs must be prepared as described in Chapters III and IV. Intricate tables are produced as a separate unit by typewriter, word processor, or typesetting and are handled as individual inserts. Very simple tables, however, can be set as part of the text and, with care, can be included in the text stream. For more information concerning tables see Chapter IV.

Printing paper selected for a camera-ready book or journal is usually an economical uncoated sheet and is not suitable for quality halftones. If it is necessary to use halftone illustrations, the author or editor should be aware that image quality will deteriorate on uncoated paper.

When it is necessary to include a photograph, an illustration window, cut to the exact size of the halftone, should be provided in the page flat. This window is a solid red block cut from rubylith or amberlith, an art material readily available in most art departments. It should be positioned in the exact place in the text at which the illustration is to appear. This block will photograph as a clear window on the page negative and will be used by the printer as a guide to superimpose the halftone negative. For a description of the halftone process, see Chapter VII.

As can be seen, each insert requires separate handling and usually means separate negatives. Obviously, these inserts add time and expense to the process. The use of large numbers of inserts should be discouraged by the editor; the added expense of time and money is contrary to the original purpose of using camera-ready copy.

Corrections

Making corrections in text in any stage of typeset or typescript material can be a painful experience. The earlier in the process a correction is made the less expensive it will be, and the more easily and quickly it can be done. At the very latest, corrections should be made before the page flats are completed. After page flats are made corrections must be calculated to fit into the same number of lines as the previous material. This will avoid the problem of a text page that might run over or under length thereby requiring that other pages be remade to accommodate the difference.

Minor corrections, such as a few words or lines, can be done as a patch. It is important that the patch be secure because the camera copy will be handled several times as it progresses through pages to editor, to printer, to cameraman.

Any major corrections that would cause three or more patches per page should be handled by retyping the entire page. The ability to make corrections in the editing draft of the manuscript is another reason for creating camera copy in a word processor, where making corrections and creating new copy are so convenient. Corrections that must be made by the printer will be expensive and will present a risk for error unless a new proof is requested. The procedure for getting this proof from the printer is time consuming and will usually delay the printing schedule.

Publisher/Printer Responsibilities

The guidelines presented in this chapter are intended to give the author/editor a general understanding of the advantages and disadvantages of supplying camera-ready copy to the printer. Once the pros and cons have been considered and the decision is made to use the camera-ready method of publication, it is important to get format instructions from the publisher and copy specifications from the printer.

Instructions and forms to be provided to author

The quality of the finished product will depend on the effectiveness of the guidelines provided by the publisher and/or the printer. The author or editor must be given specific instructions which cover everything on the printed page, including size of columns and margins, and also must include limitations.

Most printers and publishers provide page flats preprinted in nonreproducible light blue ink, PMS 310 or Capeco 544. These forms outline in blue the area for the text on the page. The text should be typed within the blue lines and should not run over them at the right or bottom margins. There are also forms that have a light blue background. Corrections can be made in the typing on these forms with white correction fluid. It is important that the preprinting be done in light blue because this is the only color that will not photograph.

The position of the journal article or book chapter title varies from publication to publication, and the amount of space to be allowed from the top blue line must be specified. The style of the titles can show variation by using bold typescript or can be set in type and a proof inserted in the text.

Authors must be told how to distinguish between right- and left-hand pages, that is, the first article in a journal or the first chapter in a book will begin on a right-hand page, but subsequent ones may be placed on whichever page comes next. Right-hand pages carry odd folios (page numbers) and left-hand pages carry even folios. The position of the folio on the first page is usually 1/8 inch below the bottom blue line.

Folios are not inserted until the entire issue or book has been typed. This avoids renumbering. Until the folios are ready to be typed, they should be penciled in lightly outside the blue lines.

Running heads can be typed with page numbers. To prevent switching running heads if pages must be added or deleted, the same running head can be used for both right- and left-hand pages.

If all of the material is not available at the time of typing, blank pages should be inserted to ensure proper page sequence.

Handling various kinds of page material

FRONT MATTER

The half title page (i), title (iii), and copyright (iv) pages carry "blind" folios, that is, they are not printed on the published page. The numbers on the rest of the preliminary pages are most often set by the printer because typewriters generally do not have type bold enough or large enough for these pages.

If there is a dedication, it should be centered on page v, which is also blind.

The foreword should be a camera-ready right-hand page and it carries a page number.

The acknowledgments, list of contributors, and table of contents follow the same system and carry page numbers.

INDEX

The index is generally typed in a double-column format after all of the page numbers have been assigned. To do this, the outline on the sheets is split vertically in the middle with a blue pencil line. The left-hand column is filled, typing main entries flush left and indenting overruns and subentries. The right-hand column is typed flush to the inserted blue line.

Processing camera-ready copy

Material submitted to the printer should be clearly identified as camera-ready copy, especially if there is a mix of copy to be typeset and camera-ready copy. These pages are fragile and vulnerable to damage through careless handling, and should be treated as original artwork. For instructions on secure packaging and shipping to the printer, use the guidelines described in Chapter IX.

Copyright Transfer

Whether copy is prepared on a computer or on a word processor or typewriter, copyright rules and regulations apply to all camera-ready copy in the same way as they do to material set in the conventional method. For further details concerning copyright see Chapter XI.

LITERATURE CITED

IFSEA, CIBA, ELSE: Model Guidelines for the Preparation of Camera-ready Typescripts by Authors and Typists. London, Ciba Foundation, 1980

Kleper ML: The Illustrated Handbook of Desktop Publishing and Typesetting. Blue Ridge Summit, PA, Tab Books, 1987

CHAPTER VII
CONTINUOUS TONE PHOTOGRAPHS AND HALFTONE PRINTING

The author, in selecting a particular image for publication, is attempting to convey a message to the reader. Successful reproduction requires that the essence of that message be maintained through the intermediate steps to the printed page. It is important to understand that tone reproduction proceeds through a series of generations of images, and that in these generations there is almost always some degradation of image. Control of this degradation determines the ultimate quality of the reproduction.

Typically, the primary image in biomedical and other scientific fields is a section or sample viewed through a microscope or other enhancing device, a piece of film (for example, a radiograph or photographic slide) or a digitized image. Alternatively, a tone drawing prepared by an illustrator may be the primary image, the latter used because it can be more subjective than the other methods. A continuous tone photograph is the usual intermediate step for conveying a visual image from an original subject to the printed book.

Tone reproduction in the publishing process consists of capturing the essential message of an original image containing gradations of density values, transmitting this message through an intermediate photograph, converting those densities to corresponding values of ink on the printed page. This conversion from continuous tone to dots of ink is called the halftone process.

Critical to making, selecting, and using quality continuous-tone illustrations is that authors, artists, photographers, editors, and printers learn a little bit about one another's craft. Some standardization of terminology and close communication about such things as objectives for the illustration process, specifications (possible limitations), deadlines, and cost will enable all members of the production team to make their best contribution to the ultimate quality of the printed illustration. The strength of the final image is heavily and sometimes solely dependent on the printing process. At times, though, one or more members of the team will take extra steps to overcome known limitations in the reproduction cycle; this should be performed with the knowledge of the author.

This chapter describes the methods used in tone reproduction and the standards required for quality reproduction.

Preparation Standards—Photographs

Why Photographs?

Photographs excel at showing specific and actual data. They should be used to show actual specimens or sites, micrographs, case histories, and any subject with complex but vital subject detail. Data generated by technical processes such as x-rays or electrophoresis gels, or data displayed on CRT screens or monitors, can be photographed directly for publication.

Photographs have an added benefit to an author. They are usually accepted by the reader as factual proof, as true reporting of information. A drawing may be questioned, the reader feeling that the artist may have adjusted some details to fit reality to the theory being presented in the text. A photo, however, is usually accepted by the reader at face value.

This credibility should not be abused by using misleading or overly manipulated photographs. Retouching should be used only to eliminate dust marks, scratches, or similar imperfections in the print for publication. Beware, too, of excessive "dodging" or "burning" (lightening or darkening selected areas of a print during enlargement). When unusual lighting or selective color filtering is used in photographing a sample, this should be noted in the figure legend so that the photograph will not be misinterpreted.

Original Photography

The preferred method of preparing a black-and-white tone illustration for publication is to make a black-and-white negative on camera original film and then a black-and-white print. Resolution of subject detail is greater and control of subject tonal range is easier than when any intermediate steps are added.

Studio or Location Photography

The photographer should be aware that the work is intended for publication in order to ensure that the contrast range is not too great, that midtone contrast is sufficient so that when the reproduction process is complete vital detail will be retained, and that the subject will be separated from the background. Furthermore, if the photograph is to be printed on high quality coated paper stock with high-resolution halftone screens using the offset process, a full-range print, with delicate or subtle detail from highlights to shadows, is possible. If the illustration is for a scientific journal, the customary reproduction technique involves 133-

line to 200-line halftone screening with a single impression of ink. Use of this technique calls for a flatter, lower-contrast range of lighting than is used for more expensive techniques of double-impression printing, or for very fine screens.

Photographic Print Quality

Image size

The information contained in the illustration must be clearly evident in print, because the author cannot discuss the picture as would be possible during a lecture. Although graphic composition is largely done with the camera, it can be modified by printing or enlarging a selected portion of the image. For scientific illustration, the main area of interest should be located near the center of the frame. Distracting or misleading details should be eliminated, but enough details to convey the full import of the illustration should be retained. If the illustration is large, more supporting details can be included. Reproduction size is a major consideration when fine details are involved. If too great a reduction is used, fine details may occupy too small an area of halftone dots to be legible.

Judging legibility of print detail at three or four feet is a good practice when critical detail is present, because it gives some idea of how the illustration will appear when reduced. To accomplish this, some layout artists use a negative (reducing) hand lens for minifying the original copy, enabling them to determine whether the image will "carry" at the smaller scale contemplated for reproduction.

Tonal range

A good glossy print can exceed the tonal range (maximum density minus minimum density) of most printing techniques (see Fig. VII–1). If a good halftone printing encompasses about 50 to 1 range (150-line screen and single impression), the photographic negative must be made to print onto a low contrast grade photographic paper, as the range of this paper most closely matches the printing process (about 100 to 1). If medium contrast or even high contrast photographic papers are used for continuous-tone prints, the tonal range is considerably beyond that of the reproduction process. In such a case, the cameraman in the printer's shop can use a variety of techniques to compress the image to suit the printing process. Because the photographer has the greatest degree of control of the process, is closest to the original, and is often in the direct employ of the author, tone manipulation should be done at or before the print stage. Hence, the continuous-tone photographic print for publication should usually have less contrast than the print for viewing. In technical terms, a full range exhibition print might have a maximum

FIG. VII–I: *Kodak Reflection Density Guide. If possible, photographers should produce prints with a density range close to that of the printed reproduction. If 150-line screens are being used by the printer, the density range of the photograph should not exceed about 1.7; otherwise, the printer will take steps to compress the tonal scale. Shown here is the photographic step table from the Kodak Professional Dataguide, Publication R-28. Holes have been punched in each density patch. Alternatively, a Kodak Reflection Density Guide, Publication Q-16, can be used. (Reproduced, with permission, from Duke University, Division of Audio Visual Education, Durham, NC.)*

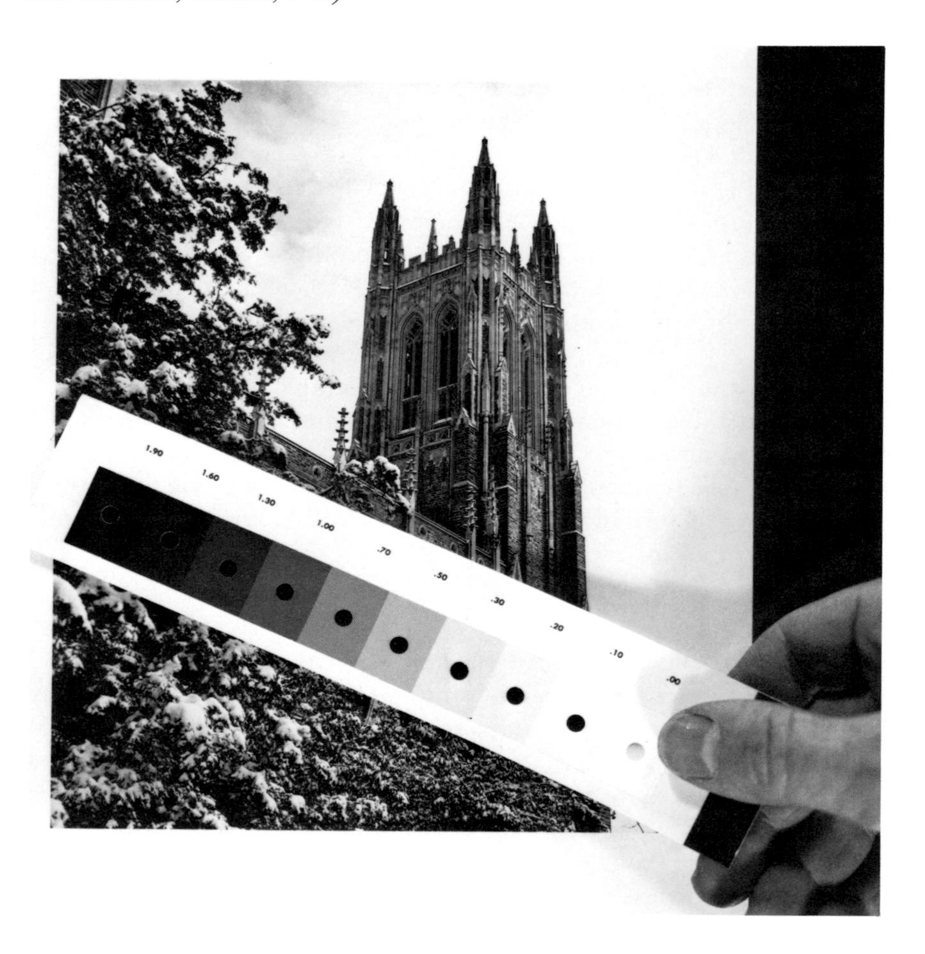

density of 1.7 to 2.25 and a minimum density of 0.10 to 0.25, or a density range of 1.6 to 2.0. A good print for average reproduction will have the same minimum density but a maximum density of only about 1.5 to 1.6. To see what these densities look like, a reflection densitometer can be used. If this is not available, the Kodak Reflection Density scale is helpful.

Making a print from a photograph is critical. Light and dark areas should both show subject detail elements. Paper contrast and exposures should be chosen to accomplish this. However, when the detail in one end or the other of the density scale is more important, that end can be favored in making the print. Chalky light areas and flat black dark areas should be avoided. The print should exhibit the full 100 to 1 range of tones, *but only at the extreme limits of the image*, that is, spectacular highlights on the subject should be printed pure white. White areas should present very light gray detail. Any small edge lines or crevices in dark areas should be printed to a deep black, but the main dark areas should retain dark gray gradations. The best criterion for middle tones is that they must "look natural," but not with as much contrast (sometimes called "crispness" or "brilliance") as would be needed in a print for exhibition in a gallery.

Some Limitations of Photography

Some subjects are very difficult to photograph with enough quality for publication. Very small subjects can be difficult because of "depth of field" limitations. The smaller the subject, the less depth or thickness of the subject is in focus at the camera's film plane. Flat or very thin subjects are usually not a problem, but round, deep or thick subjects are a problem in macro- or microscopic photography and may require specialized equipment or techniques, such as scanning and transmission electron microscopy. In extreme cases, a drawing may show the subject more accurately.

Clear or colorless subjects may need special staining, special filtering, unusual lighting, specialized equipment, or a combination of these, to bring out subject detail. Microscopic subjects, such as diatoms or tissue sections, are good examples of this. On a larger scale, an animal such as a jellyfish would call for such techniques.

Some very complex subjects may require visual simplification for the key details to show clearly. Framing the subject closely and eliminating distracting background or non-essential details are often effective. Although lighting can be manipulated to highlight only the most important features, this is not always practical. Some subjects in this category may be better illustrated by drawings, as the artist can omit extraneous detail and emphasize important elements far more easily than can the photographer.

Photographic Papers

Glossy surface papers (F surface) should be used, as they can carry more detail than matte surfaces. Medium- or double-weight paper is preferable so that the print can be handled with fewer chances of bending or cracking. Papers with white highlights and neutral black tones should always be selected. Full development should always be employed, as curtailment of development introduces a number of other variables, the primary problem being little possibility of producing a duplicate print.

It was once believed that ferrotyping of prints for reproduction was necessary; this was primarily a matter of protecting the print and providing more "crispness." The graphic arts camera did not "see" any difference between a ferrotyped and a non-ferrotyped print, although the ferrotyped print darkened somewhat because of the heat involved in the ferrotyping process.

The question of making prints on a fiber-paper base or a resin-coated base is a question not of image quality but rather of archival storage and cost. Resin-coated papers, introduced in the early 1970's, are rated as having "commercial" keeping quality, that is, more than 20 years. For true archival quality prints, fiber-based photographic papers should be used in conjunction with strict processing, washing, and storage requirements. Many automated photographic laboratories have gone exclusively to black and white printing on resin-coated papers and add a considerable surcharge for hand processing of fiber-based papers. For scientific illustration, retouching must be supervised by the author, to ensure that only flaws are repaired and that no spurious information is introduced. Retouching is best accomplished by attaching an acetate or tissue overlay to the print so that the author can mark any dust or scratch marks for retouching.

Prints of color slides

Because much of the scientific communication revolves around teaching with 2 × 2-inch slides, the original image for black-and-white reproduction may be a color slide. When publication is contemplated, separate sets of slides for black and white and for color should be prepared at the outset. If this is not possible, intermediate negatives should be made on a panchromatic film, preferably of 4 × 5 inches, with exposure and processing techniques customized to the slide (see Fig. VII-2). If the slide is a full-range slide, contrast reduction similar to that described below for radiographs should be used. Common faults in prints from color slides include high-contrast prints devoid of highlight or shadow detail, or of both, and prints with highlights that have grayed considerably.

CRT Photography

Cathode-ray tubes (CRTS) are increasingly being used as output devices in medical and scientific recording. Data are digitized and stored on

FIG. VII-2: *Black-and-white reproductions of color slides are frequently used as scientific illustrations because of the prevalence of the 2 × 2 inch (35 mm) teaching slide. Unless photographers take steps to produce a low-contrast negative and suitable print (A), a high-contrast print lacking highlight, shadow detail, or both will result (B). (Reproduced, with permission, from Duke University, Division of Audio Visual Education, Durham, NC.)*

A

B

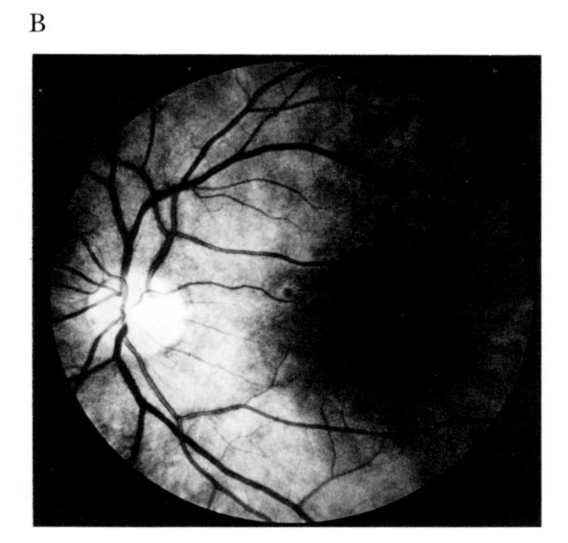

magnetic tape. In many cases, computers are used for analysis or enhancement. The typical CRT image is not difficult to photograph and print for publication. If possible, a film should be exposed by the technician on the normal output device of the recording apparatus, as this device contains a much higher quality CRT than the one used for viewing. The CRT should be adjusted according to the manufacturer's specifications. Once the film is obtained, an intermediate negative and print can be made by normal means, but with the understanding that CRT film images generally do not require the contrast reduction methods needed for conventional radiographs or the contrast enhancement methods used in low-contrast situations.

If the CRT image is to be photographed, a normal contrast film should be exposed at a shutter speed of 1/8 second or longer (focal plane shutter) in a darkened room. Faster shutter speeds (shorter exposures) may record less than a full image, yielding a diagonal band, and a room light will degrade the normal contrast of the image. The camera should be positioned parallel to the face of the CRT and sufficient depth of field should be used to capture the curvature of the CRT in sharp focus from edge to edge. Prints that contain a pattern, dots, or grids may cause an

FIG. VII–3: *Moiré pattern. A. Normal photograph shot through halftone screen. B. Wavy pattern seen in sky is undesirable Moiré pattern. (Reproduced with permission of Jon Belcher, Capital City Press, Montpelier, VT.)*

A

B

undesirable moiré pattern (see Fig. VII–3) when shot through a halftone screen, creating wavy lines or an out-of-focus image in the halftone. Halftone proofs made from photographic prints that are likely to create such an image must be carefully checked in the proofing stage.

Clinical and Research Laboratory Photography

This category of subjects includes electrophoresis, microscopy, chromatography, and the many other newer tests used in the clinical and research laboratory (see Figs. VII–21–31).

Problems of quality reproduction are generally concerned with very faint bands in electrophoresis gels or with indistinct detail in other reactions. If the main details are clear the reproduction process can carry them. However, if the details are faint good reproduction will be very difficult. High contrast films, special lighting for detail separation, filters for rendering tonal contrast of colored subjects, and a cylindrical lens (a test tube filled with water or a glass rod) to sharpen electrophoresis bands are among the tools available to the photographer. For example, if detail is present in the negative, variable-contrast paper can be used so that shadows can be printed for high contrast and midtones for medium contrast.

When difficult subjects are presented for reproduction, each member of the team must try to achieve the greatest legibility on the printed page; often photographers have more flexibility than the printer. It is helpful if the author informs the photographer and editor about which details in the photograph are significant (see Fig. VII–19).

Prints of Radiographs

Conventional radiographs have a density range of 3.0. Because this scale is logarithmic, there is an intensity range of 1–1000 from density minimum to maximum. This range is best compressed to fit the printed page (density range of about 1.5) in the intermediate negative step (see Fig. VII–4). Several techniques are useful: over-exposure and under-development (techniques commonly used in other areas of photography with high contrast subjects), simple masking, double-masking, and electronic scanning.

Making good prints of radiographs is difficult because the tonal range of conventional radiographs is often 1000 to 1. Furthermore, densities in a radiograph, for example, a normal chest x-ray, must fall within a certain range, because a density outside that range may indicate a pathological condition. There may be faint detail or density gradations in both the heavy- and light-density areas, often requiring labeling so that other radiologists can follow the discussion. Another problem is that images from conventional radiographs and most of the newer imaging methods of computed tomography (CT), ultrasonography, and nuclear medicine are on film and are viewed by transmission rather than reflec-

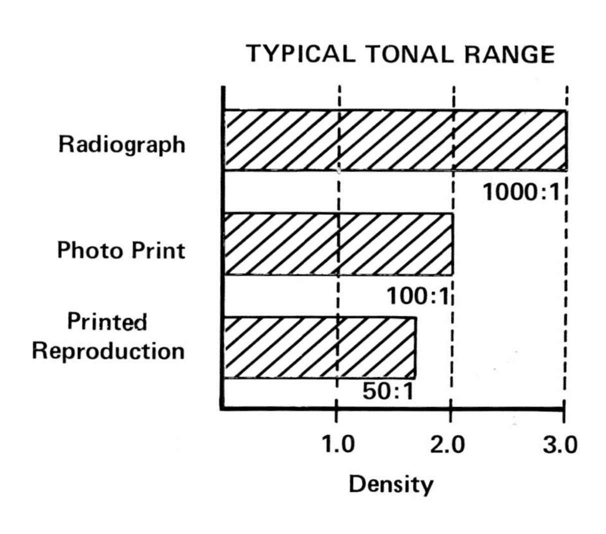

FIG. VII–4: *Basic to successful reproduction of continuous tone illustrations is the fact that the printing process, by its very nature, cannot yield the fidelity of the original or, sometimes, even of a photographic print of the original. Shown here are the typical tonal ranges of a radiograph, a photoprint, and a printed illustration. (Reproduced, with permission, from Duke University, Division of Audio Visual Education, Durham, NC.)*

tion. Photographic papers and the printed page are viewed by reflected light and do not have the tonal range of film.

Over-exposure and under-development (see Fig. VII–5) make possible a reasonably careful control of tone compression; if large-format film is used, compression can be individually tailored to each radiograph.

Simple masking includes a variety of techniques—many small, switchable light bulbs; paper cutouts; pencil shading; and diffuse film density. These techniques are limited in the degree of compression they offer and they tend to be somewhat mechanical and therefore uneven.

Double masking offers maximum flexibility with very natural results. This is a manual process which involves several hours of labor for each radiograph. The techniques of over-exposure and under-development and the technique of electronic scanning, provide results almost as good as double masking, but with much less time required. For all but the most difficult radiographs, double masking is not necessary.

Electronic scanning is done with a device made by the LogEtronic Company. It senses high- and low-density areas of the radiograph, and

FIG. VII–5: *Internegatives of a radiograph. Over-exposure and under-development were used to condense the range of densities.*
A. Internegative showing normal contrast and density range.
B. Internegative for the same radiograph with density and contrast reduced to "fit" the printed reproduction process. (Reproduced, with permission, from Eastman Kodak Company, Rochester, New York.)

A

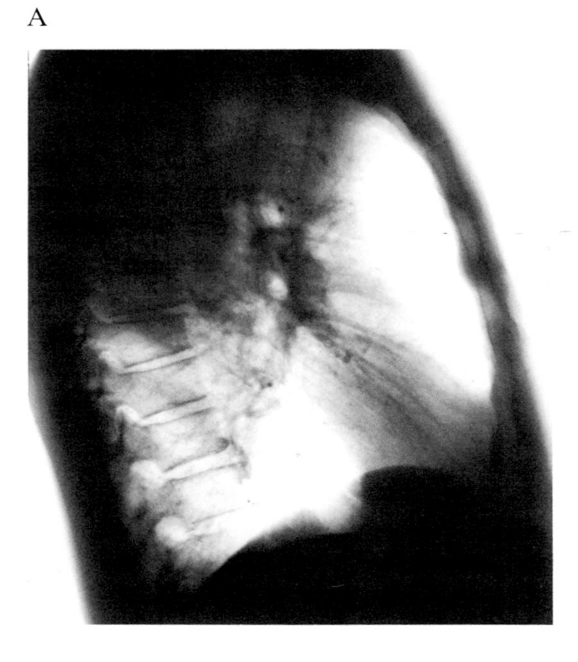

B

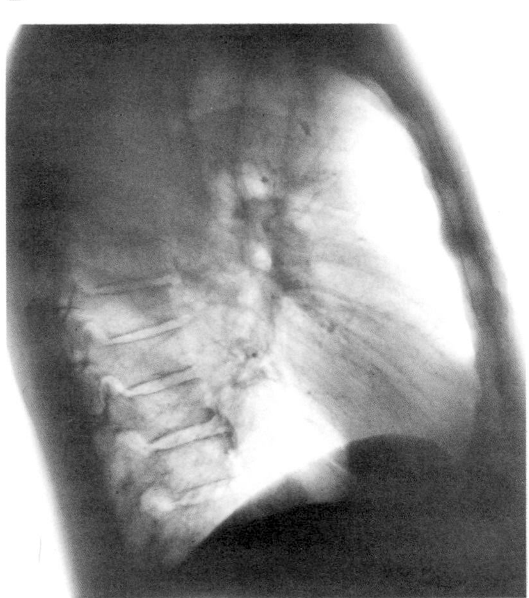

by variation of either intensity or velocity makes the high and low densities of the radiograph "look" closer together during the intermediate negative exposure. This machine is capable of a considerable range of compression, and the intermediate negative can then be altered in processing to provide even more modification (see Fig. VII–6).

If scanning is too severe, the resulting image appears "unnatural" to some observers. With all but the most difficult radiographs, electronic scanning accomplishes only slightly more than could be achieved with carefully controlled over-exposure and under-development, and not all authors have access to the LogEtronic scanner.

FIG. VII–6: *Electronic scanning of radiographs greatly facilitates making intermediate negatives and prints that are ideally suited for reproduction. As illustrated here (lower left), a low-contrast developer further aids in producing a negative that is filled with detail yet has the proper contrast and tonal range for reproduction. (Reproduced, with permission, from Eastman Kodak Company, Rochester, New York.)*

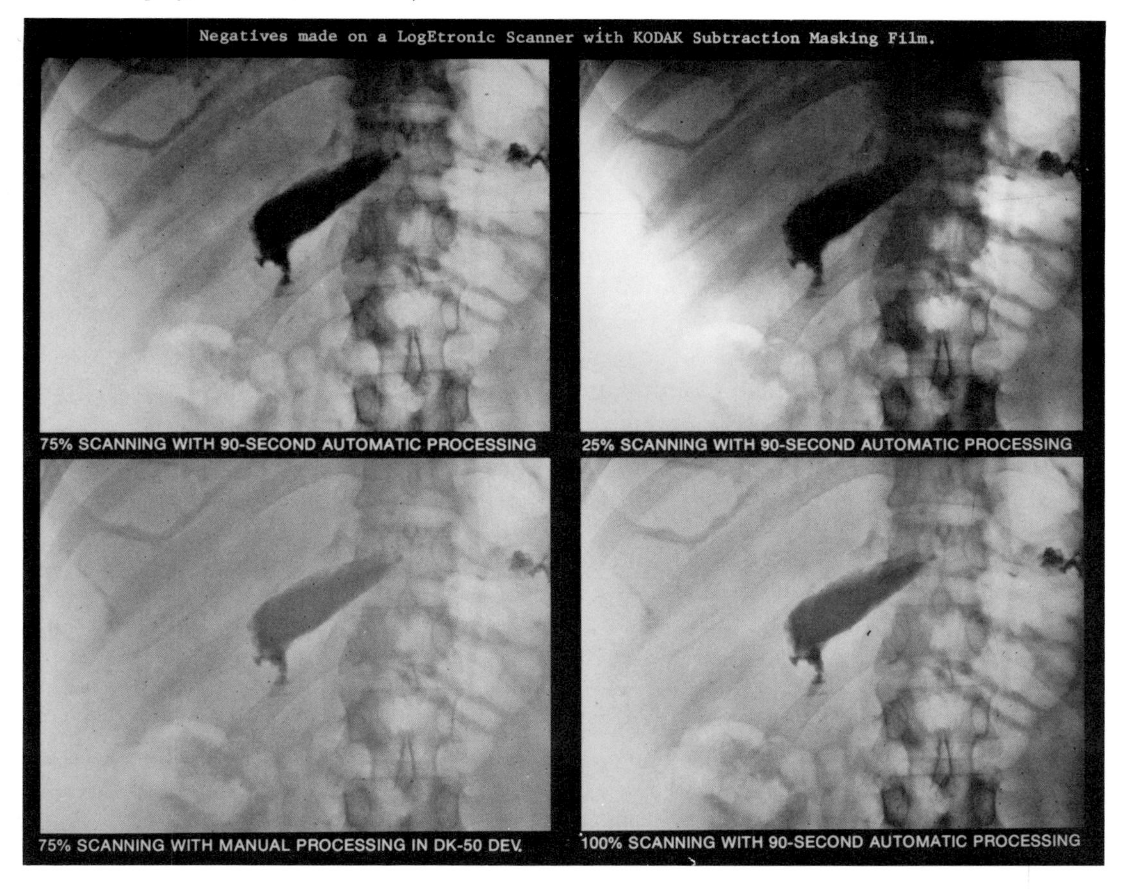

Publication Requirements

Publication requirements for tone illustrations reflect the evolution of the discipline they serve and the technologies that have developed around the image they must convey. Some represent good practices in all scientific disciplines, some have been made obsolete by new publishing technologies, and some are unique to the authors' specialized fields. It is difficult to mandate requirements for all scientific publications, and such mandates are beyond the scope of this section. At the outset, authors should consult specific journals or editors for instructions to the authors, and should peruse previous issues for a perspective. What follows is intended to be a general guideline of good practice.

The Planning Process

As soon as an idea for an article or book begins to take shape, the author should begin to coordinate development of the illustrations and the text. With this approach no words will be wasted, and the illustrations will perfectly complement the text, clearly conveying the author's ideas and discoveries to the reader.

Planning ahead is particularly important for the photographs. When a series of experiments begins, with the intent of publication, publication-quality photos can be planned for and taken during the appropriate stages of the proceedings. If photographs are considered only after the procedure is concluded, it may be difficult or impossible to reconstruct key elements for the camera.

Early in the planning process, the author should consult with the photographer (or illustrator) who will provide the illustrative material for the article, considering the optimal method for communicating key information and setting up a schedule for production of artwork and photographs, with enough time allowed for quality results.

The very best black-and-white prints for reproduction are made by directly photographing the original subject material with black-and-white film, from which the final print is made. Each step that intervenes between the publication print and the original subject degrades the image, with progressive loss of detail and clarity. For example, if color slides are needed for a lecture and a paper will be published in black and white on the same subject, take both color slides and black and white negatives from the same subject on the same occasion. Black-and-white prints from the negatives will be approximately equal in quality to the image on the color slide, but if the color slide is rephotographed for black and white some image degradation will occur, and defects present in the slide (for example, from the grain of the slide or dust and scratches in the emulsion of the slide) will appear in the black-and-white print. Similarly, it is better to rephotograph the subject (if still available) for a new original print than to copy an existing print of the same subject.

Republishing tone material that has been already published with the halftone process creates special problems and should be avoided if at all possible. In many cases, the original prints can be obtained by contacting the author or publisher of the work.

When discussing particular photographs to be made for an article or book with a photographer, be sure to provide the following information: 1) that the photographs are for publication; 2) which publication (provide a sample copy, if you have one); 3) any particular "instructions to authors" having to do with artwork or photographs; and 4) key points or details of each photograph to be taken (or printed from your negatives), and how you want them presented. Set a mutually agreed-on date for completion of the work. Be sure you have discussed and decided what needs to be done, and leave a phone number where you can be reached if questions or problems arise.

Many types of scientific images are photographed and published. If there is a question about preparing a specific image for publication, the author or photographer should look at similar published images in journals or books. This will provide standard views or conventions.

Size and Number

Photographs submitted for halftone reproduction should be no larger than 8 × 10 inches (20 × 25 cm) so that they will fit in standard files and can be easily mailed. They should not be smaller than 5 × 7 inches (13 × 18 cm) for ease of handling and viewing. Size is important, not only for ease of handling in an editorial office or at a publisher/printer but for quality of reproduction. The general rule is that the closer in size the reproduction is to the original, the better the fidelity. Journals often have size conventions for particular kinds of illustrations; for example, on a two-column page a brain scan may consistently be printed in half-column width. Familiarity with the way a particular journal presents its illustrations, keeping in mind the percentage of reduction, will greatly facilitate choosing the size of the original print. Problems can arise if there are too many illustrations in relation to the length of the accompanying manuscript. Layouts with figures continuing for pages after the text stops are awkward. In addition, some publications have a set maximum number of illustrations permitted per article, and charge the author for any additional figures.

Multiple sets of illustrations are usually submitted with an article for publication (consult the appropriate Instructions to the Author); copies of the manuscript and sets of illustrations are usually sent simultane-

FIG. VII–7: *Cropping and layout. A and B are originally submitted photographs of cranial CT scans. There is unnecessary information in A and distracting machine "noise" in A and B. Also, B is at a different magnification than A. C and D are the same two pictures as they appeared in print. Unwanted parts of the images have been cropped out or covered with Amberlith. D has been reduced more than C so that the images are now of the same magnification. (Reproduced, with permission, from Lufkin RB, Wong WS, Winter J, Callisen HH: Low cost digital teleradiology systems. AJR 140:379, 1983.)*

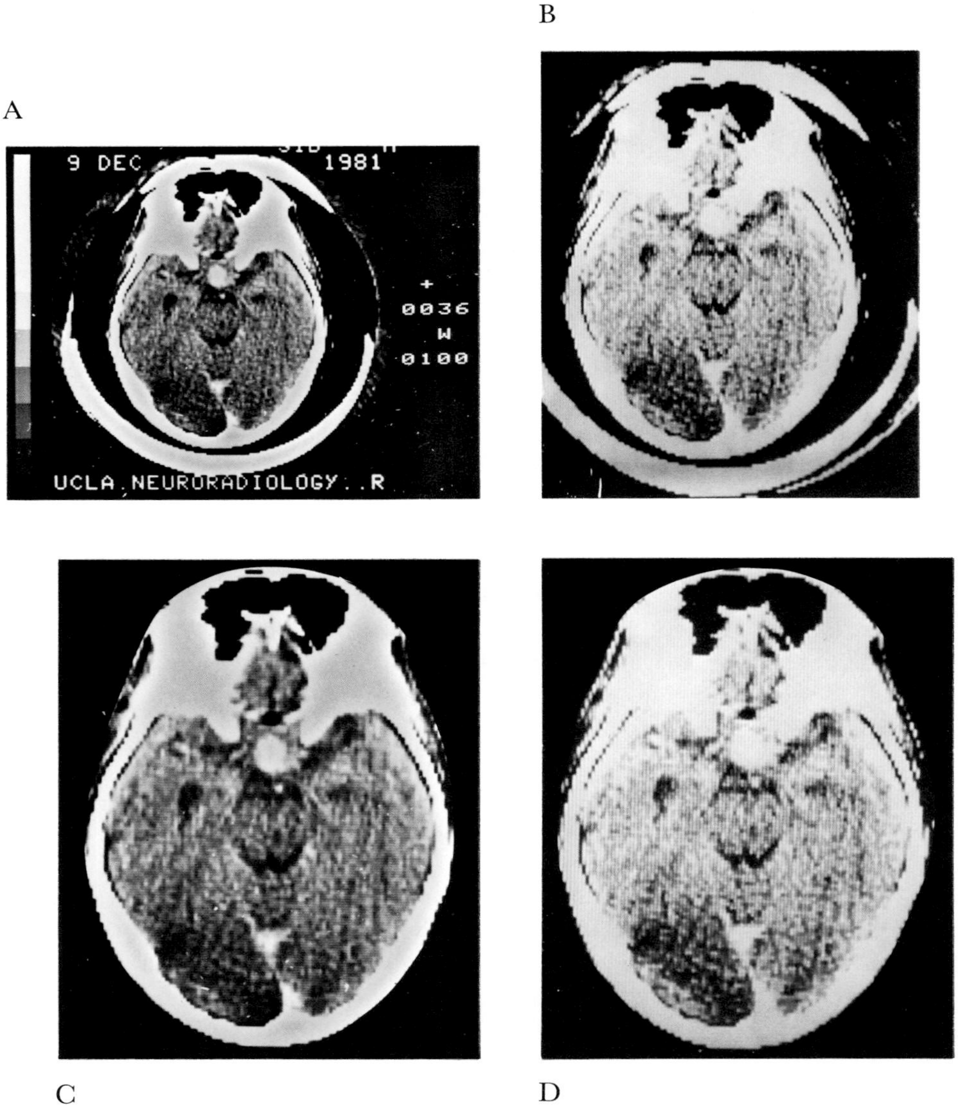

B

A

C

D

ously to two or more reviewers. Complete sets of figures should be packaged separately, such as in different envelopes. The sets should never be paper-clipped together, as the clips damage the prints. Separation of sets of figures is especially important for stereoscopic illustrations, which involve pairs of identical illustrations that can be confusing if they are not kept in order. It is particularly important that the author keep a full set of the illustrations. The author should have photographic prints made of any original image (an artist's drawing, a radiograph, CRT, an instant-camera print). Any copying or labeling should be done on the copy, not on the original, in case more copies of the unaltered original are needed later.

Cropping

Never remove the white border around photographs submitted for publication unless the journal specifically requests trimmed-and-combined prints. These borders may be needed to allow the editorial office to put crop marks on the picture without damaging the image. The images themselves should contain only the essential material, and should be large enough that the subject matter can be easily identified. Authors can indicate preferred cropping on a tissue or acetate overlay. Editorial office staff have enough knowledge of the subject that their cropping does not delete areas of interest. Particular care should be taken in making sure that crop marks create 90° corners; if a print is cropped askew, it will take time and money at the printer to "square up" the crop marks. This does not mean that the crop marks must be at right angles to the edge of the paper, just that the cropped image must be a square or a rectangle.

Cropping is done for several reasons. A picture submitted by an author may need to be "squared up." Prints of several sizes may be cropped to similar size so that they all can be photographed by the printer at the same percentage of reduction (see Cost Controls, page 183). Often only parts of prints are selected for publication, for example, using only one half of a chest radiograph to show abnormality in one lung in detail. Photographs of histologic specimens are often cropped so that they will not be reduced in printing to the point that the cells are difficult to differentiate.

Identification

A label, sometimes supplied by the publisher or printer, should be affixed to the top or bottom of the figure according to the publisher's instructions, to orient the figure. The information generally includes identification of the article and the figure number. This information, along with the first author's name (unless the publication requires author anonymity for reviewers), should be typed on a self-adhesive label. Writing directly on the photograph with pencil or ballpoint pen damages the image. Identification written with a felt-tipped pen on nonabsorbent photographic paper can transfer to the image of the picture underneath in the stack, or bleed through to the image side. Bits of paper with the figure number and other information should never be paperclipped to the photograph, as the clips can damage the print surface and create wavy distortions that may be visible after reproduction.

Labeling

The most common method of labeling is to use transfer (press-on) letters and arrows (such as Letraset, Chartpak). Style and size of labels and arrows should be the same in all figures for a given manuscript. Ideally, sizes should be consistent in the finished, printed work, so larger labels should be used on pictures that will be reduced more than others. Labels, arrows, and other internal identification must be explained in the figure legends.

Grouping and Mounting

Pictures can be grouped, such as parts A through D of Fig. VII–7, for several reasons. Grouping can be done by case, subject matter, or other combinations of pictures with something in common (see Fig. VII–8 and Orientation and layout, page 181). In determining the size and layout of illustrations, the author should crop (cut) and combine (tape together) prints into a multipart figure. If it is an editorial office's responsibility to design the layout of multipart figures the author may want to include a diagram or a cut-and-combined set of prints (in addition to uncropped prints) to indicate the preferred layout. These will also be used for review purposes. If the printer is expected to crop and compose multipart figures, he should receive uncropped prints and detailed instructions about size and figure layout desired, including a diagram showing the placement of the figure parts (see Orientation and layout, page 181). The printer may prefer not to have figure parts taped together if the various parts differ in density and therefore will require separate shooting. For example, if several medical photographs are to be printed as one figure, body parts could be the same size to facilitate comparisons. Unless the publisher (or journal) specifies mounting procedures, prints should not be mounted, for example, glued to sheets of paper inside acetate sleeves or on heavy cardboard. These methods may delay handling and increase mailing costs. Heavy cardboard mounting also can cause inaccurate reproduction; for example, if the edge of part A of one figure overlaps the edge of part B, the two parts will be different distances from the camera lens. In addition, some camera work done by printers requires that the photograph be sufficiently flexible to be wrapped around a drum. Finally, heavy cardboard-mounted photographs may not fit in the vacuum frame of a camera copyboard.

Overlays

Overlays are very useful to indicate preferred cropping, position of labels, and especially important parts of images (so that detail can be matched exactly or enhanced). A tissue-paper overlay can be used with the markings made in *soft* pencil (so the photograph is not damaged), although the opacity of the paper can hinder precise positioning of labels. Acetate overlays are transparent, allowing direct visualization of cropping and placement of labels or arrows, and they are easily marked with grease pencil. If professional-quality labels are applied to an acetate overlay, the picture can be reproduced as a composite; the photograph is shot with a halftone (dot) screen, the overlay is treated as non-screened linework, and the two are combined. This eliminates the fuzzy edges on letters and arrows produced by the dot pattern in the halftone screen, but is much more expensive than doing a single reproduction of a labeled photograph.

Orientation and layout

Certain conventions are observed in the way some illustrations are printed and viewed. For example, an ordinary chest x-ray (radiograph) is viewed as if the reader were facing the patient. Some other common orientations are presented in the section on Primary Images (see page 201). These decisions regarding orientation are made in the editorial office. The printer cannot be expected to identify all the different types of images and know how they should be oriented, so the editorial office must give clear instructions if the submitted picture requires special handling (for example, to flop a picture right-for-left).

Layout of illustrations is important in presenting the material to the reader. Usually the editorial office or publisher decides how illustrations fit into general page design, such as always putting pictures at the tops of pages. However, there are many other ways of designing figure layout for emphasis or consistency. Often one figure is a group or series of several photographs with something in common. Some examples are: 1) case: several photographs of a single patient; 2) time: a series of prints showing change over specific time periods (see Fig. VII–8), such as in one joint with gout; 3) diagnosis: photographs showing the same kind of abnormality in several patients, or one disorder as depicted on several imaging methods; 4) sequence: especially with cross-sectional scanning, a series of contiguous images; and 5) position: examples of a structure on photographs taken at different angles or with a patient in different positions (for example, prone, supine, lateral, etc.). Care should be taken in combining pictures for the best effect; for example, if a knee joint is to be compared in several side-by-side pictures, the joint should be at the same level in all pictures so that the reader can scan from one picture to the next without having to search for the region of interest.

If all or many pictures in an article are of the same type, they can be made the same size by cropping and/or reducing. If they appear next to each other in columns or on facing pages, they can thus lend symmetry to the page layout.

When pictures will be grouped or a special layout is wanted, detailed instructions must be given to the printer. It is easy to say that each figure should be one column wide. However, complicated layouts require special attention from the editor and the printer. The editorial office can specify that illustrations be made certain widths (for example, make figure one 20 picas or 8.5 cm wide, make figure two 29.5 picas or 12.5 cm

FIG. VII–8: *Radiograph of gout involving foot joints. This is an example of a series of similar images obtained at various time intervals in a single patient. B was taken 13 years after A; C was taken 14 years after B. Magnification and cropping of all three images are similar, to facilitate comparison. (Reproduced, with permission, from Bloch C, Herman G, Yu TF: A radiologic reevaluation of gout: a study of 2000 patients. AJR 134:784, 1980.)*

A B C

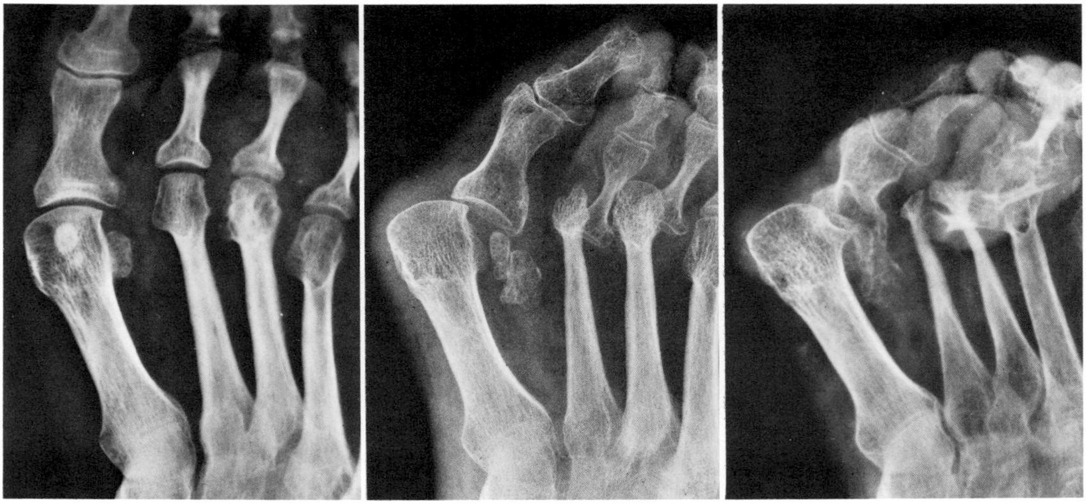

wide), or they can determine, using a proportional scale (also known as a reduction wheel), the reproduction percentage (for example, shoot figure 1A at 66%, shoot figure 1B at 73%). However, the reproduction percentages must *match* the desired widths; that is, if instructions are to shoot at 55% (85 mm wide) and the printer determines that shooting at 55% will make the picture 97 mm wide, he may change the percentage, change the cropping, or contact the editorial office to clarify the discrepancy. All of this causes delay, possible error, and additional cost. Layout instructions to the printer should include sketches of pictures grouped into figures (so they will know, for example, that 1B goes beside 1A, and 1C goes under 1A) and any special requests, such as: *flop* figure 1B right-for-left; place figures 1 and 2 on facing pages; reproduce figure 3 lighter than copy. Printers should be instructed about tooling between multiple illustrations in a plate.

Cost controls

Costs of reproducing illustrations can vary greatly, depending on how much special handling is needed and how many printer's negatives are required. A publication may limit the number of illustrations permitted for a single article, charging the author for any additional pictures. Pictures that are grouped as a single figure can be photographed already combined at far less cost than if each picture is photographed separately (see Fig. VII–8–9). Combined shooting requires that all pictures photographed at once be shot at the same percentage of reproduction. However, if two pictures are to be grouped and reproduced at the same percentage, but one is much denser (darker tones) than the other, it may well be worth the expense of having them done separately, as quality can suffer if the printer averages the densities to shoot them together.

Corrections of photographs done in an editorial office often cost less than the same corrections done by a printer (for example, adding Amberlith, airbrushing) (see Fig. VII–7), if the printer adds a handling charge or overhead expenses. Labels/arrows applied to a clear overlay and shot as linework separately from the halftone will be better looking (see Overlays, page 181) but much more expensive than photographing a labeled print. When unsatisfactory proofs and pictures are to be sent back to the printer to be redone, the possible improvement must be weighed against the added cost and time.

A high-technology printing plant requires large investments in delicate and expensive machinery to produce elegantly printed products in the shortest possible time. In the scientific/biomedical disciplines, the rush to publish places great financial pressures and time constraints on the printer.

Thus far, we have addressed the "ideal" halftone. Now we will address cost-saving methods. The principal rule is, *what saves the printer time reduces*

FIG. VII–9: *Averaging versus masking for best fidelity. Inset may be shot separately so densities are not averaged. (Reproduced with permission of Rockefeller University Press, from Westermark B, Porter KR: Hormonally induced changes in the cytoskeleton of human thyroid cells in culture. J Cell Biol 94:42–50, 1982.) Gray scale at right indicates tonal values and readings for density values from highlight to shadow, thus providing a "thermometer" to judge densities in the halftone. (Portions of figure 9 also used in figures 11, 12, 13, 14 and 18.)*

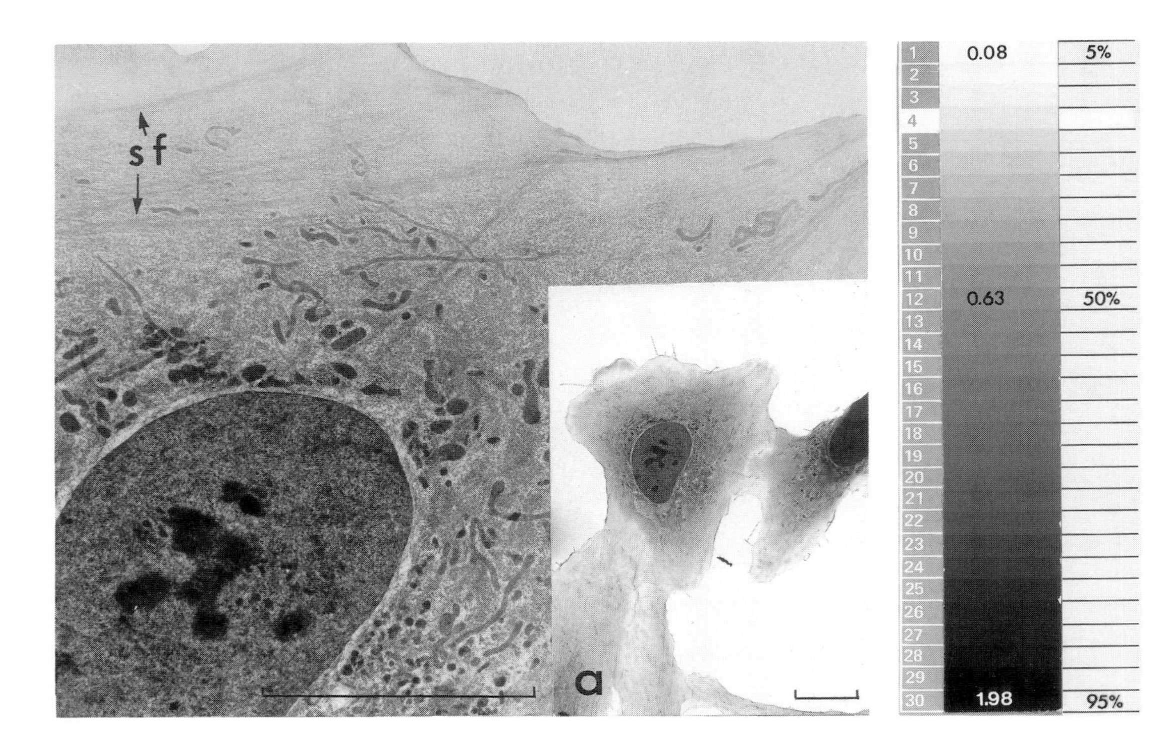

schedules and costs. The printer's efficiency is directly proportional to the quality of the copy received and the accuracy and timeliness of the publisher's instructions.

With regard to halftones, the publisher should consider a number of cost-saving possibilities. Sometimes halftones that require equal magnification or reduction and which have very similar density and tonal value can be placed on the copy board and shot together. This process of multiple or "gang" shooting reduces camera time and saves money. Tonal values in each photograph must be very close or the individual halftones will suffer from loss of quality.

In shooting sequential or chronological layouts—again, where tonal values are close—a multipart figure may be shot as one halftone. In cases where density or tonal value differ significantly, certain subjects within a plate may be masked and shot separately. Presentation of multiple subjects as one figure will reduce the publisher's costs. If an illustration containing eight subjects is shot separately, the cost will be eight times greater than that of shooting the illustration as a single exposure. In shooting a single shot, however, the density values across the illustration will be averaged throughout the eight subjects.

Halftones

A halftone is the printer's approximation of the photographer's photograph or the illustrator's tone illustration. Because the printer's end product is ink on paper and because his raw ingredients are white paper and 100% black ink (for clarity's sake we shall discuss tone only in terms of black ink, but the same basic rules apply to any color ink), the printer creates the illusion of continuous tone by recreating the image of the photograph in groupings of dots of ink. The size and closeness of these dots determines the grayness or blackness of the image and enables the printer to create a tonal range of grays from a single value of black ink.

Breaking up of the image into dots is accomplished by placing a halftone screen between the image (photograph) and a piece of unexposed film in the printer's copy camera (see Fig. VII-16). The film is then exposed to the photographic image through the camera lens and the halftone screen. When the film is developed, the result is a grouping of small dots recreating (in negative form) the photographic image. In the offset printing process the negative is exposed to a light-sensitive anodized aluminum plate. Ink adheres only to the dots thus exposed. The plate is put on the press and ink is applied. The image is transferred to a rubber blanket and then to paper. When the ink makes contact with the paper, the dots approximate the original photographic image. There are certain limitations to the fidelity of this re-creation, which are discussed on page 197.

Each tone illustration has three broad areas of density known as highlight, shadow, and middletone. Areas of the image containing more white space than ink are referred to as highlight areas; parts of the illustration with more ink than white space are called shadow areas. The areas between black and white are referred to as the midtone, or middle tone range.

The primary or first generation image is the author's raw material. The second generation image is usually created by a photographer and results in a print. This second generation image, which we will call original photo or "copy," must first be "read" for its density values so that the proper exposure of photo to halftone film can be determined. This reading determines the number of exposures, the duration and intensity of the light source, and the direction of light sources. In some camera rooms, density values are determined by a visual comparison with a control gray scale. More sophisticated systems use photoelectric cells or densitometers which give digital readouts of gray scales. Recent innovations combine densitometers with computers to replace value judgments otherwise exercised by skilled cameramen.

After density determinations are made, the proper lens and lighting adjustments are set, the copy camera is loaded with film and halftone screen, the cameraman exposes the halftone film to the subject, and the halftone negative is developed.

Typically, a quality halftone negative employs three exposures of the film; a main, a flash, and a bump. The main is the principal image-creating exposure. High-intensity light aimed towards the copy board reflects light off the original photo through the camera lens and the halftone screen onto the film. The flash is made by removing the copy and again exposing the film to light (no image) through the halftone screen. This extends the range of the halftone by increasing the number of tones of gray. The bump is made by removing the halftone screen and exposing the halftone film to the original photo. This compresses the tone in a particular range, thus giving greater visual contrast. A proper balance of these three exposures is necessary to achieve the best halftone. Too much flash will flatten the tones. Too much bump will obscure detail.

Halftone Screens

Selection of the proper screen is essential to faithful reproduction. A good practice is to test the effect of a variety of screens on the type of primary images commonly found in the discipline that is the subject of publication. The correct screen for reproducing electromicrographs may be inappropriate for CT scans. Screens that yield good results for halftones printed on sheet fed presses may print poorly on web presses. Some knowledge of screen characteristics will assist the editor or publisher in guiding the printer to the desired effect.

A halftone screen is a continuous tone sheet of film on which a grid has been superimposed. Normally, the grid is ruled at a 45° angle (the angle at which the dot pattern is least discernible to the human eye.) The distance between the lines of the grid determines the number of dots per inch. These lines consist of rows of dots. The screen itself is an emulsion of silver salts which allow various degrees of light intensity to pass through it. The intensity is governed by the original photograph. The white (highlights) of the photograph will reflect more light intensity than the gray midtones or darker shadows.

The finer the grid the more dots per inch and therefore the more image detail. Grids, or screen size, are measured in lines per inch and range from 85 line screen (newspapers) to 300 line screen. Commonly available screen sizes are 133, 150, 175, 200, and 300 (see Fig. VII–11). Higher-lined screens are more difficult to print. Fine screen rulings require higher quality papers and inks and more stringent press conditions.

Another factor in the selection of screens is the shape of the dot. In

FIG. VII–10: *Halftone screen superimposed on original photograph to illustrate elliptical dot screen pattern. (Reproduced with permission of Jon Belcher, Capital City Press, Montpelier, VT.)*

FIG. VII–11: *Screen size. Lines per inch. This figure shot at same size as original (50% dot in 0.52 density). (Reproduced with permission. See figure 9.)*

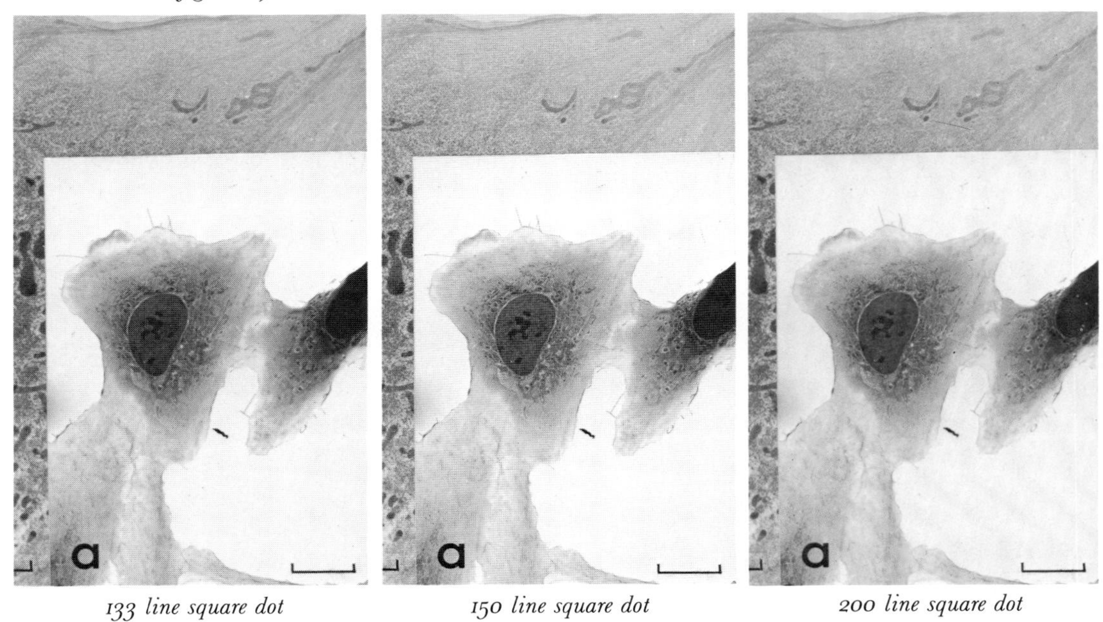

133 line square dot *150 line square dot* *200 line square dot*

FIG. VII–12: *Shape of halftone dot. All figures enlarged 300% from original. 50% dot in 0.52 density. All screens 133-line. (Reproduced with permission. See figure 9.)*

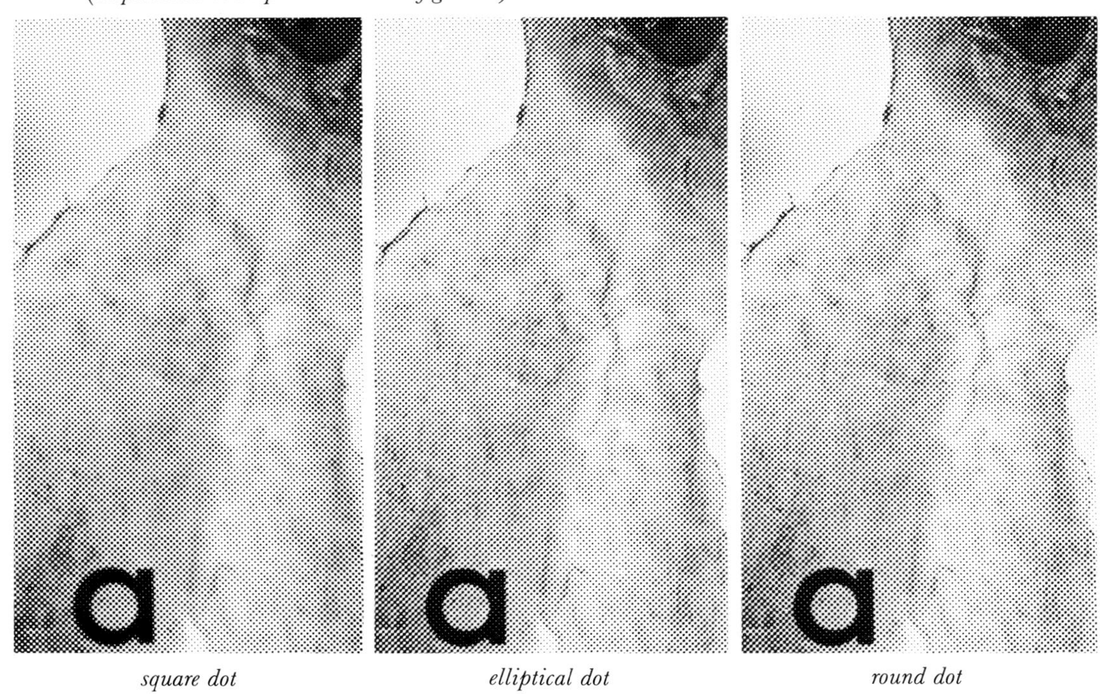

square dot *elliptical dot* *round dot*

FIG. VII–13: *Shape of dot. All figures enlarged 900% from original. 50% dot in 0.52 density. All screens 133-line. (Reproduced with permission. See figure 9.)*

square dot *elliptical dot* *round dot*

FIG. VII–14: *Shape of dot. All figures enlarged 900% from original. 50% dot in 0.52 density. Dual-dot 150/212-line screen. (Reproduced with permission. See figure 9.)*

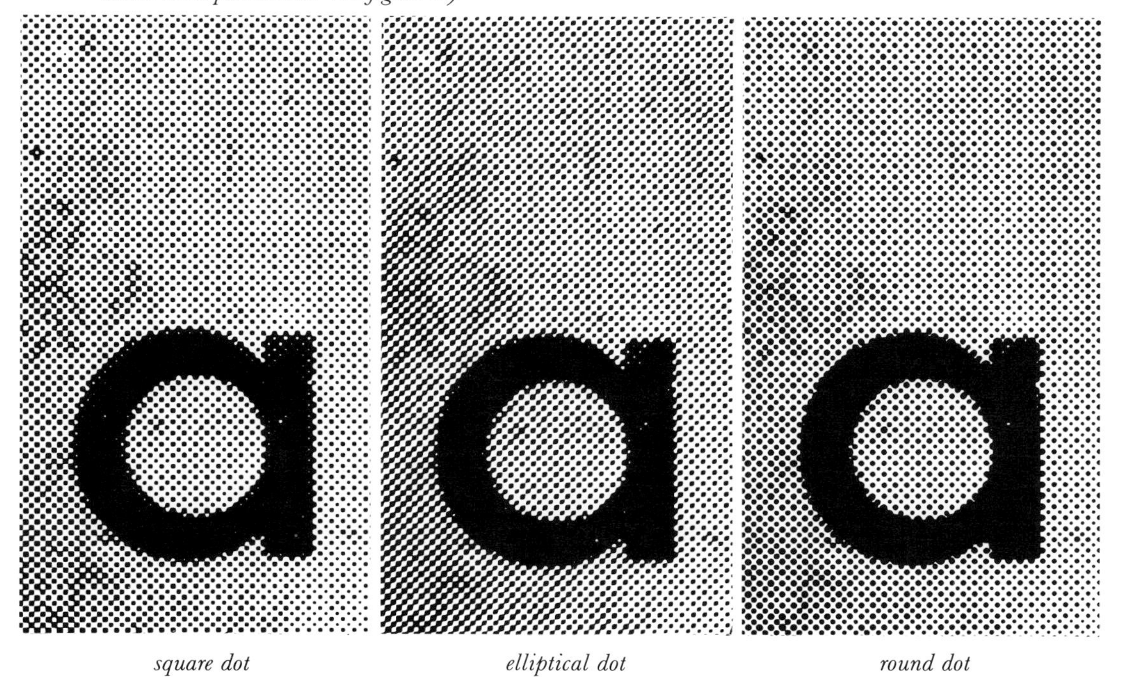

square dot *elliptical dot* *round dot*

FIG. VII–15: *Dot formation with crossline screen. (Reproduced, with permission, from Graphic Arts Technical Foundation, Lithographers Manual, Pittsburg, PA, 7th Edition: Chapter 5:53, 1983.)*

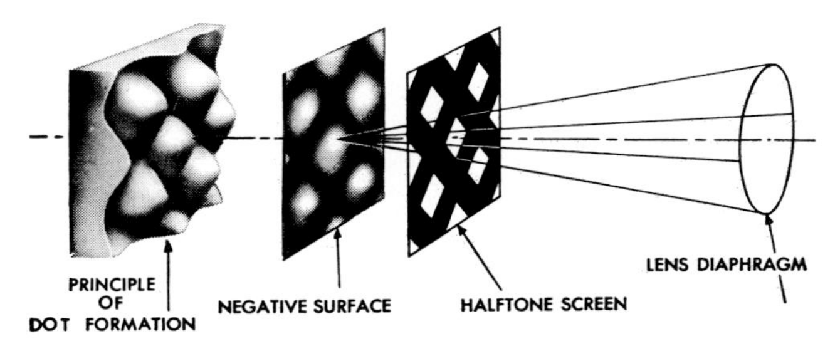

most common use are the square dot and the elliptical chain dot (see Fig. VII–12–14). Round dot screens are frequently used in web offset printing.

The square dot is the most commonly used dot form. The elliptical dot more faithfully reproduces subtle gradations in skin tones which occur in the mid-tone range. Inspection of the 50% dot will show "softer" contact of ink than is present in the square dot. Where sharper definition is required in the mid-tone range, the square dot may be preferable.

The true shape of the dot is seen in the 50% dot, that is, where the printed dot is equal in area to the unprinted part of the paper. In that particular space of the halftone the inked dot covers half of the area, and the unprinted or "white" area is the other half of the space.

The size of the screen (dot) alone does not necessarily determine the quality of the printed tone illustration. Quality depends, in addition, on the control exercised in the printer's offset preparation department, and in plate-making and printing press controls.

Camera

The essence of the production of halftones resides in the camera department of the printer's shop. Within this department all major decisions affecting the quality of reproduction of the copy provided take place. Here are melded the subject provided to the printer, the screens used, the capabilities of the camera and metering of the light, the accuracy of plate-making, the proficiency of presses and pressmen. As printing has progressed from a delicate craft to a highly refined technology these decisions have become more quantitative than qualitative. As the printing process has become more costly the need for greater cost effectiveness has imposed rigorous demands on the printer because it affects author-editor-publisher's ability to deliver to the reader the message intended at an affordable price. As a result, the copy camera has become larger, enabling more halftones to be shot at the same time. Copy cameras range in size from 8 × 10 inch to 34 × 44 inch, the dimensions referring to the maximum size film the camera can accommodate (see Fig. VII–16). Since line work requires less stringent optics, the printer employs a larger camera and lens for line work and a smaller, higher-quality camera for halftone shooting.

In the typical printing plant today, the camera occupies two rooms separated by a wall containing the lens and its mechanism. It resembles an old fashioned box camera on a larger scale. In the room in front of the lens is the copy board (a vacuum plate to hold the photograph or illustration). The lens mechanism and bellows ride on tracks to provide proper focus. Here also are the light source that provides illumination for the copy, the mechanism for monitoring the amount and direction of the light source, and the nature of the light source itself. In the room behind the lens, the cameraman loads his film and inserts his selected halftone screen in the camera back, exposes the film, removes it, and de-

FIG. VII–16: *Copy camera showing camera back, bellows, track, lights, and copyboard. (Reproduced, with permission, from NuArc Company, Inc., Niles, IL.)*

FIG. VII–17: *Original photo being placed on copy board of vertical camera for halftone exposure. (Reproduced, with permission, of Jon Belcher, Capital City Press, Montpelier, VT.)*

velops it in a red light environment, frequently with the aid of an automatic film processor.

Monochromatic Flat Bed Scanners

The most recent development in halftone reproduction is a scanning system that employs integrated cells (similar to those used in video cameras) which are sensitive to light and "read" the original illustration. This reading is digitized and fed, in the form of electronic impulses, through a computer which is capable of manipulating the information and the resulting halftone image.

Once manipulated, the electronic impulses are used to drive a helium-neon laser beam which creates the halftone dots on film which will be used to make the printing plate.

Scanners read the original photograph in very small units on a bed (image area) approximately 12 × 18 inches in size. As many as 10,000 light cells sense the image placed on the bed. Scanning of the image on the bed is very fast compared with conventional halftone shooting, especially where photos have large areas of shadows.

The integrated computer permits a number of different programs to manipulate the image, much as main, bump, and flash exposures do in conventional shooting techniques, but the scanner can move the highlight, 25%, 50%, 75%, and shadow dots to achieve greater fidelity to the original photograph.

The computer is capable of a wide variety of programs which require skillful management by the operator if the halftone is to represent the image faithfully. A danger exists that improper programming will distort the image and confuse or degrade the message the author had in mind.

Conveying the Message

Where top-quality halftones are required it is never sufficient to demand that the halftone look just like the photo, because that objective is not obtainable with available technologies. The author/editor must ensure that the offset cameraman understands the essential message of the photograph, the intrinsic purpose for its reproduction.

In lieu of that specific type of communication on each photo, it is incumbent upon printer and publisher to investigate the similarities of certain photographic disciplines (see Primary Images, page 201) and to locate the essential message in the area of the gray scale where it appears. A working knowledge of highlights and shadow, contrast, detail, and middletone, as well as the function of the gray scale, will enhance the author/editor's chances for delivering to the reader the message contained in the photograph.

The anatomy of a printed halftone consists of a series of screened dots of different sizes. In their final printed form, these dots will be smallest in the highlight (lightly shaded) areas and largest in the shadow (dark) areas of the image. Unlike a continuous tone photograph, a halftone negative contains no pure whites and no solid blacks. Consequently, a halftone has a shorter tonal range than a photograph. The halftone screen covers the entire image area of the photograph, so dots should be discernible in all areas of the printed subject. The only control is the size of the dot or, conversely, the white spaces between the dots. A high quality photograph may have a tonal range from pure white (the color of the photographic paper) to total black.

We describe this range as the photograph's density range. By density we mean light stopping ability. Printers (cameramen) use a density scale that ranges from 0 (white) to 2.4 (black). This scale is logarithmic. With the aid of light-reading densitometers, printers can determine density values of photographs and can translate them to equivalent values of percentage of ink to white space on paper. This process involves the use of reflection densitometers to read light reflected from opaque surfaces and transmission densitometers to read light passing through negatives. The offset press, however, can not achieve a range equal to that of the photograph even under the best of circumstances. The smallest dot it is capable of printing is approximately a 3% dot; the largest dot it can carry or hold open without "filling" the white space is about a 95% dot.

The average reproduction range of a typical printing press using coated paper is the difference between the unprinted white paper (0.10 on the logarithmic scale) and the solid (no space between the dots) black ink (1.55 density). Thus, average reproduction range equals approximately 1.45. Average density range of a photograph is approximately 1.80 (shadow 1.90 – 0.10 highlight). Therefore, tone compression is necessary.

Enhancement of detail in the highlight or shadow area of a halftone negative can improve the accuracy of the message transmitted to the printed halftone. This enhancement is accomplished by spreading or extending the tonal range in the message carrying area of light or shadow. To achieve this effect, the midtone (50% dot) must be moved towards that density area. If contrast is to be maintained, detail in the opposing area must be sacrificed. If the 50% dot is moved into the shadow area to create better definition, then detail is lost in the highlight area. The reverse also holds true (see Fig. VII–18).

Light reflecting off the photo onto the halftone film is the only way to transfer detail; since the bump compresses the tonal range, too much bump exposure will drop out detail. Consequently, the main exposure

FIG. VII–18: *Moving the 50% dot. 133 positive magenta elliptical dot screen (same size as original; see Fig. VII–9).*

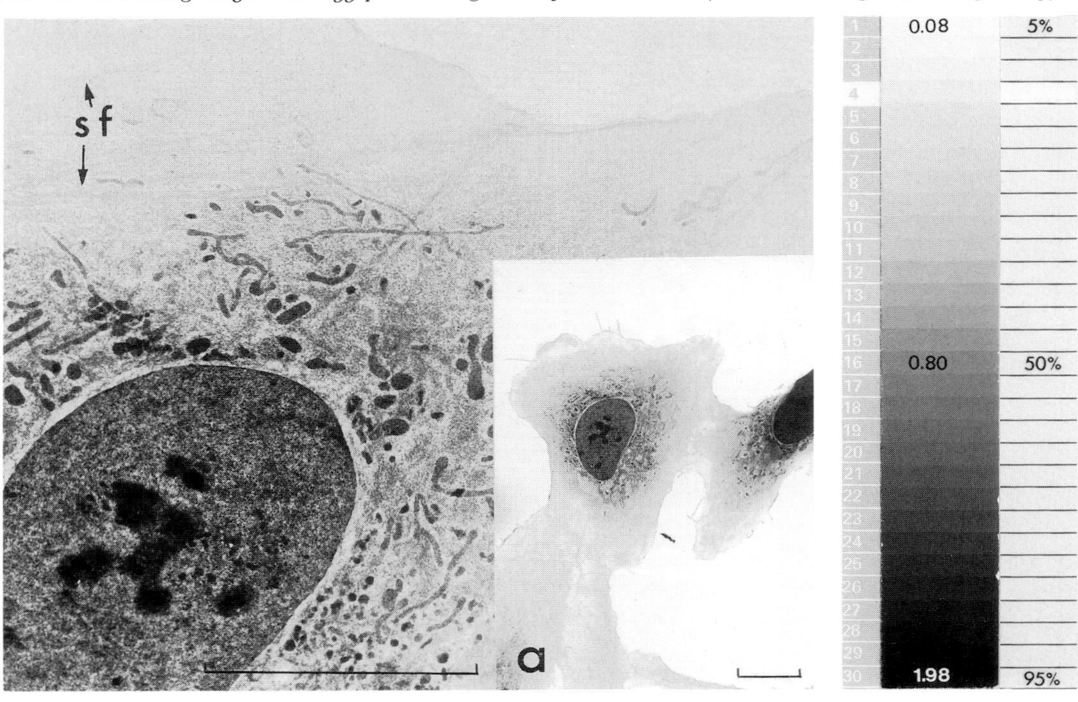

A. 50% dot in 0.80 density range.

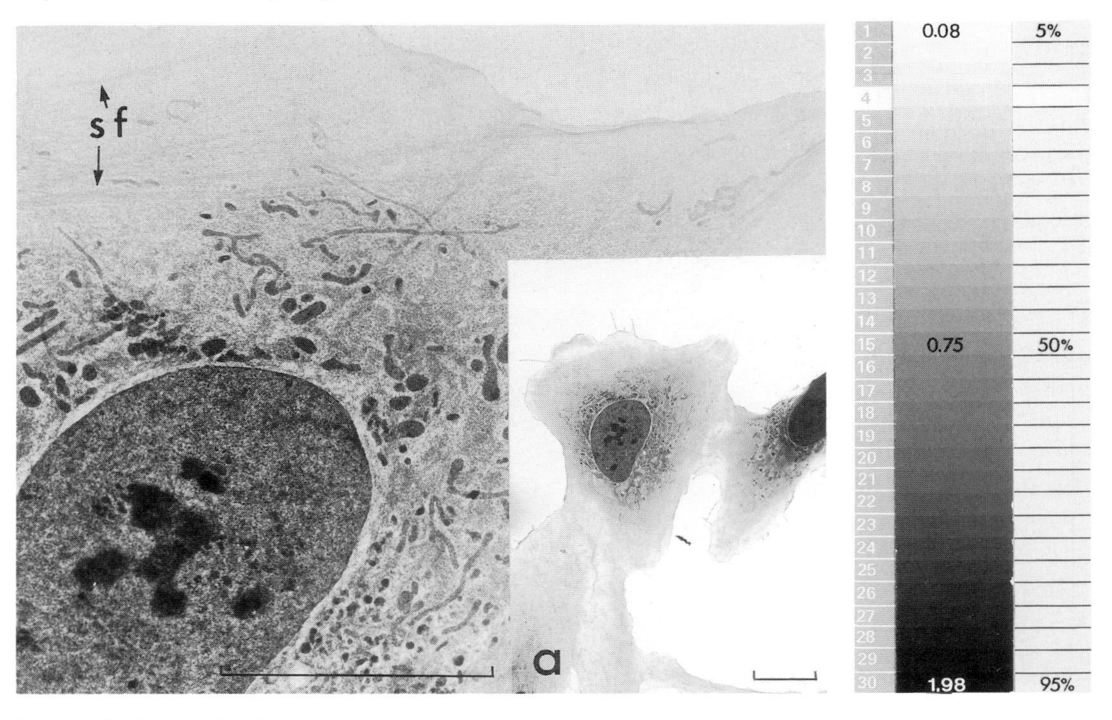

B. 50% dot in 0.75 density range.

FIG. VII–18

1 2 3 4 5 6 7 8 9 10 11 12 13 14 15 16 17 18 19 20 21 22 23 24 25 26 27 28 29 30
0.08
0.52
1.98
5%
50%
95%

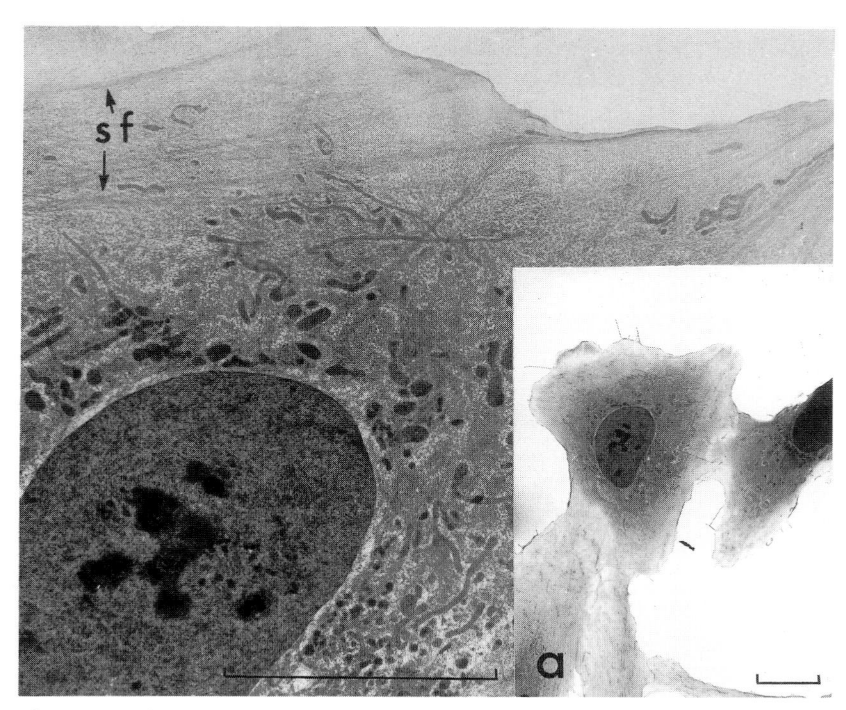

C. 50% dot in 0.52 density range.

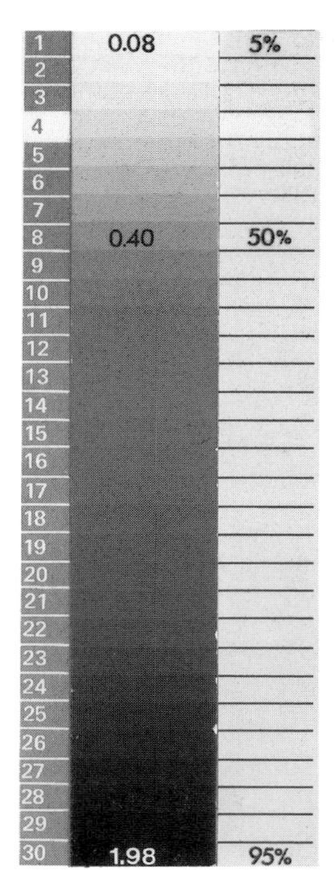

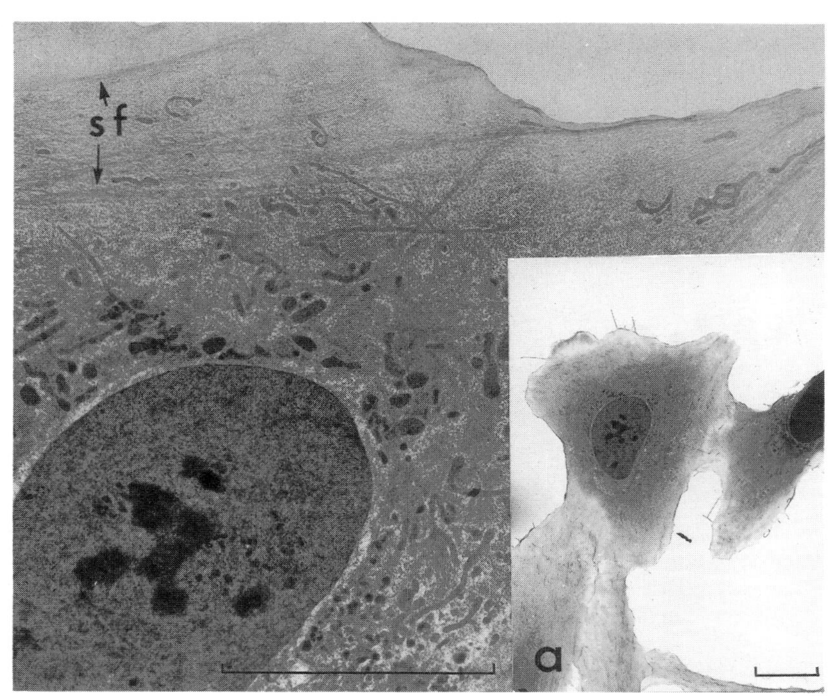

D. 50% dot in 0.40 density range. (Reproduced by permission. See figure 9.)

must generally be reduced. It is important that a screen be selected with a tonal range that most closely approximates the tonal range of the photograph or the area of the photograph that carries the message.

Halftone proofs

Once the halftone negative has been developed, the continuous tones of the photograph have been converted to dots for press printing. This is the appropriate time to check the halftone to ensure that the message has survived the transition. Changes in halftones in steps subsequent to this one become prohibitively expensive and time consuming and, from a practical standpoint, sometimes impossible.

Three types of proofs are available from most printers for viewing by author/editors. These are, in order of increasing accuracy and expense: the loose blue or Dylux®, the photoprint, and the press proof. Each of these proofs may be known by a variety of names (see Glossary, Chapter XII).

A printer's chief consideration in recommending a proof to an editor should be how faithfully it will represent the finished (printed) halftone. A publisher's or editor's choice may ultimately be determined by price. The press proof is the most accurate, since it duplicates the actual press process. For all but the most exacting requirements, its price makes it prohibitive. Press proofs are usually printed on the same paper as the production run but on a smaller press than a production press, so even this proof is not an exact match to the production-run halftone.

A loose blue is the least expensive and least precise. It is a contact print made from the halftone negative under ultraviolet light in daylight conditions on a pre-sensitized paper. Blue, brown, or gray in color, it lacks the high resolution or contrast to be expected in a published high quality halftone, and is seldom relied on for more than positioning in page proof. Some editors with experience in viewing loose blues find them sufficient for content. More often, some type of photoprint is selected.

Like a blue, a photoprint is contacted from the halftone negative but is of much higher quality, made in the darkroom and produced on high quality photographic paper. In simplest terms, it is a quality photographic print of the dots on the halftone negative. It shows detail that is lost in the blueline proof. Under carefully controlled conditions, a printer may produce a more exacting photoprint than press-printed halftone. Limitations of the press and differences between photographic and printing paper account for this; the consequences must be guarded against by printer and author/editor alike. The message of the original image, clearly visible in the photoprint, may be lost in the printed halftone. Alternatively, a poor-quality photoprint may lack detail or contrast which may reappear in the press-printed version. Photoprints which sometimes look flat or lack contrast may be enhanced by the characteristics of ink and press sheet. Communication among printer, publisher, and editor should resolve any serious deviation between photoproof and printed halftone.

Regardless of the type of proof, it should be reviewed by author and editor for detail and contrast. Unsatisfactory halftone negatives should be reshot before stripping is done, to avoid unnecessary expense.

Dot Gain

The halftone negative, as its name implies, is the reverse of the press-printed halftone positive. The original (halftone negative) dot has a tendency to alter its shape from camera negative to printed halftone. This characteristic is called dot gain.

Halation occurs in the making of the halftone negative and may be likened to a halo around the moon with a fuzzy outer concentric ring. The nature of the screen is such that light is concentrated in the center of the dot with less reaching the perimeter. The result is a blurred effect which will be seriously enlarged in the printed halftone unless measures are taken to "tighten" or "harden" the dot. A halated dot is frequently the cause of inferior halftones. Dots are best viewed through a 12- to 30-power loupe or microscope, but considerable education about the visual characteristics of dot structure and its pathology is required before such inspection by author or editor is constructive.

Distortions of the photographic dot can be caused by the following conditions:

(1) pressure — the printing plate mounted on the press cylinder contacts the rubber blanket transferring the image to it and, in turn, from blanket to paper; the pressure on these relatively soft surfaces enlarges the dot;
(2) paper — surface, whiteness, and opacity may alter the shape or affect the viewer's perception of the dot;
(3) ink — the consistency, speed of drying, and thickness of the ink film may cause the dot to spread;
(4) slur — improper movement within the press itself can elongate the dot;
(5) doubling — movement of the blanket will create a tail on the dot.

Optical dot gain can result from improper paper selection. Lack of opacity or clarity of reflection can cause the eye to see a shadow which will register in densitometer readings. Smaller dots appear to gain more than larger ones, but in actuality they do not. The amount of gain is the same on all second generation dots. In the outer perimeter width, dot gain is the same on all dots. The midtone, having the greater perimeter or area, will gain the most.

Dot gain is expressed as a percentage of enlargement from the original dot to the press-printed dot. Some causes of dot distortion can be virtually eliminated. Careful monitoring of darkroom processing solutions, press inks, blankets, and consistent use of the same printing paper reduce the variables that affect dot gain or distortion. Any change in specifications from halftone film to printing paper (including press room temperature and humidity) is apt to influence the dot. It is necessary that the printer compensate for those conditions that cannot consistently be controlled.

Compensation for dot gain should be exercised in the camera department when the halftone negative is shot. In litho presses, dot gain can run from 5% to 15% in the mid-tone. Some cases of dot gain of up to 40% have been found in newspaper printing. Failure to compensate for and control dot gain results in unsatisfactory halftones.

Improving Halftones

As noted above, the shape of the dot produced by the halftone screen, size of the grid, accuracy of density readings, color of screen, etc., influence the quality of halftone reproduction. Some other techniques are available to printers to enhance halftones or more faithfully reproduce original photos. These include moving the 50% dot into the highlight or shadow area, use of a combination or dual screen, and use of double-dot (duotone black) screens.

This process may involve balancing all three exposures (main, flash, bump) and may require complicated mathematical calculations and/or plotting of graphs by experienced cameramen. Density readings, dot area percentages, screen calibration, dot gain, printing quality, and paper selection must all be taken into account for photographs. Sophisticated densitometers/computers now take the guesswork and calculations out of the darkroom, *but only if the author/editor informs the cameraman about the part of the density range of the photo which contains the essential message.* Tissue or acetate overlays prepared by the author are useful in aiding the cameraman (see Fig. VII–19).

Dual dot screen

The dual screen has two sets of dot formations. Whereas a single screen has a single screen ruling, the dual screen has a large dot in the midtone to shadow area, a small dot in the highlight, and a combination of the two in the middle tones to shadows. This screen provides for a very fine dot where highlight detail is important and for double-dot formations

FIG. VII–19: *Scientific illustrations may not have the area of interest in the highlight area (A). In fact, sometimes subject details are located in the shadow or maximum density areas (B). It is helpful to editors, publishers, and printers for the author to note the areas of importance so that the reproduction process can be faithful to the intent of the illustration. These notations should be made on an acetate overlay, as in the figure below. (Reproduced, with permission, from Eastman Kodak Company, Rochester, New York. Overlay prepared by Thomas Hurtgen, Duke University, Durham, NC.)*

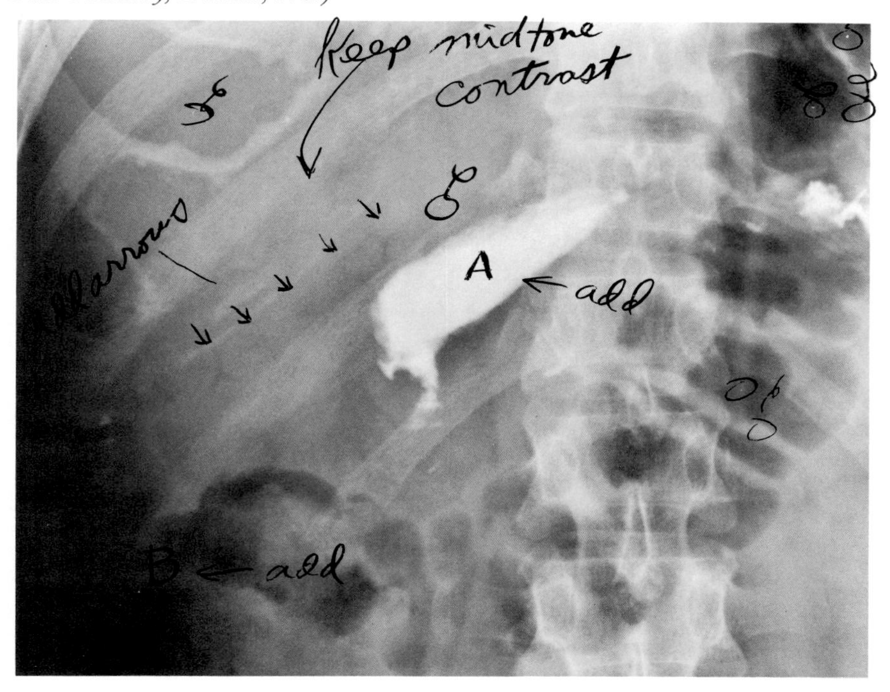

in the midtones to shadow (see Fig. VII–14). A dual dot should be used only where specific shadow detail is important and where press, paper, and ink are ideal for control of double-dot formations.

Duotone

The duotone uses two separate halftone negatives that are shot at an angle of 45° and 75° each. Each halftone is exposed separately to the copy and produces its own negative. Two plates and two passes on the press are required. Sometimes two identical black inks are used, sometimes a brown tone or other color. Because duotones require two sets of plates and two printing passes, they are substantially more expensive than a standard single screen or dual dot halftone. A properly prepared

FIG. VII–20: *Comparison of conventional halfone using one halftone screen (upper pair) and a duotone using two halftone screens (lower pair). Note the enhanced shadows and highlights in the duotone. A. Habit of* Cavendishia martii *(Luteyn JL and Lebron-Luteyn ML, 6364). Peru, Ayacucho, 50–54 km NNE of Tambo, 2200–2560 M, 3 Dec. 1978. B. Close-up of influorescence of* Cavendishia punctata *(Luteyn JL and Lebron-Luteyn ML, 5427). Peru, Junin, Tarma-San Ramon Rd., Chuquisaca-Huacapistana area, km 72–78 above Rio Palca, 1560–2000 M, 29 Feb. 1978. (Reproduced, with permission, from Luteyn JL, Flora Neotropica Monograph 35, Bronx, NY, The New York Botanical Garden, 1983.)*

and printed duotone produces a rich, warm, three-dimensional effect with wide tonal range and excellent detail (see Fig. VII–20).

Ink and paper

Since the final halftone reproduction, that which the reader sees, consists of ink on paper, these two ingredients are of paramount importance to a high-quality halftone. Paper is a major cost factor in the publishing process. Its selection directly influences the price of the publication.

High-quality halftones require costly papers. Factors influencing high

FIG. VII–21: *Radiograph. A. (Reproduced, with permission, from Fisher MR, Mintzer RA, Rogers LF, Lin, P-JP, Bova JB, Doerner SM: Evaluation of a new mobile automatic exposure control device. AJR, 139:1055–1059, 1982.) B. (Reproduced, with permission, from Chang VP, Benjamin R, Jaffe N, Wallace S, Ayala AG, Murray J, Charnsangavej C, Soo CS: Radiographic and angiographic changes in osteosarcoma after intraarterial chemotherapy. AJR 139:1065–1069, 1982.) C. (Reproduced, with permission, from Wagner ML, Singleton EB, Egan ME: Digital subtraction angiography in children. AJR 140:127–133, 1983.)*

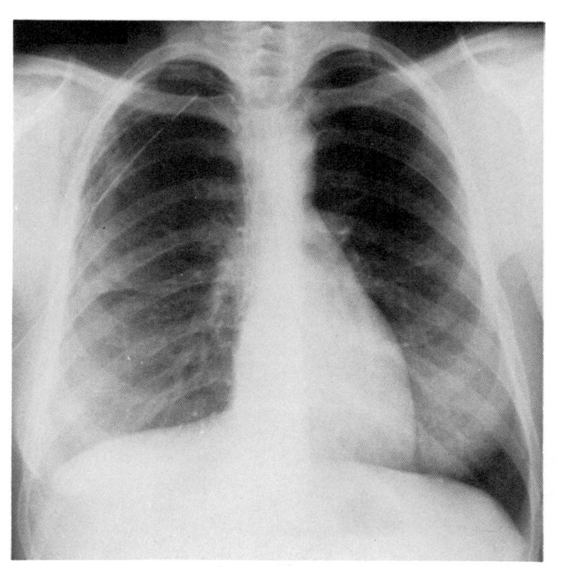

A

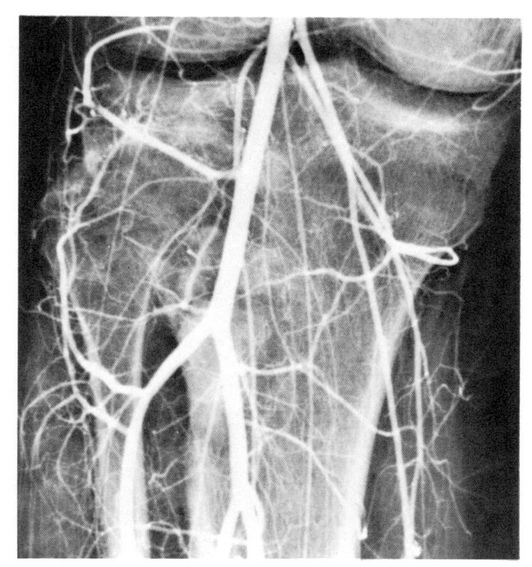

B

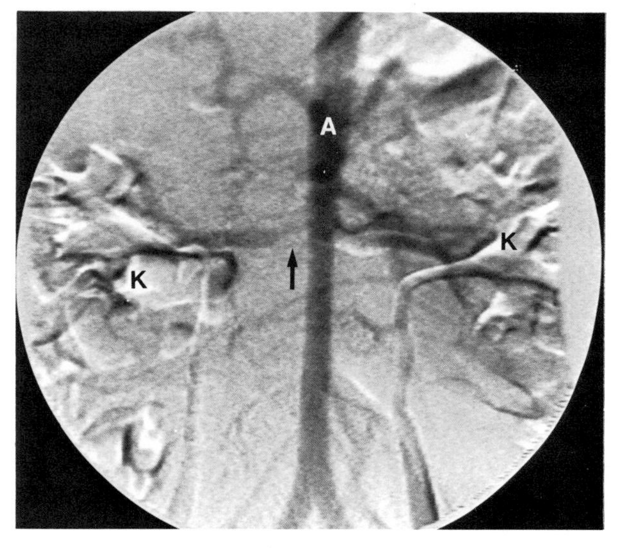

C

FIG. VII–22: *Computerized tomogram (CT scan). A. (Reproduced, with permission, from Libshitz HI, Jing BS, Wallace S, Logothetis CJ: Sterilized metastases: a diagnostic and therapeutic dilemma. AJR 140:15–19, 1983.) B. (Reproduced with permission of Dr. Michael Brant-Zawadzki, University of California, San Francisco.)*

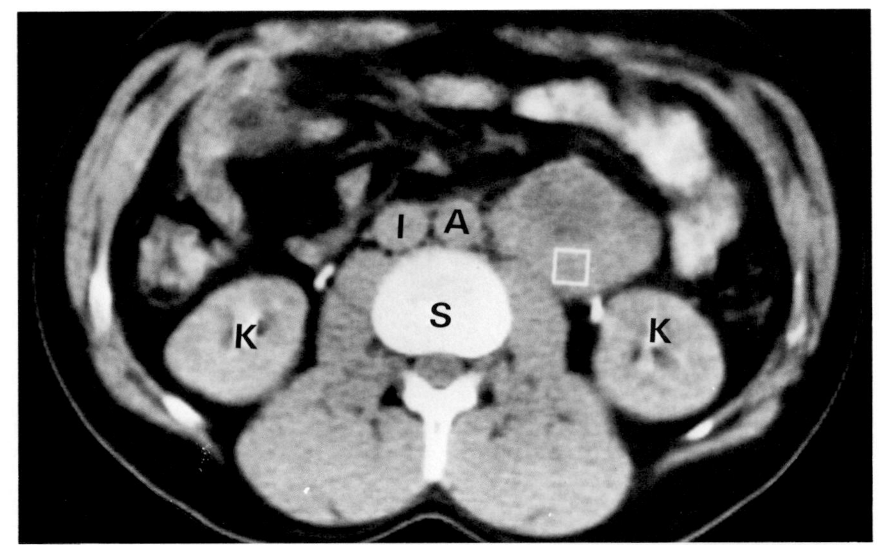

A

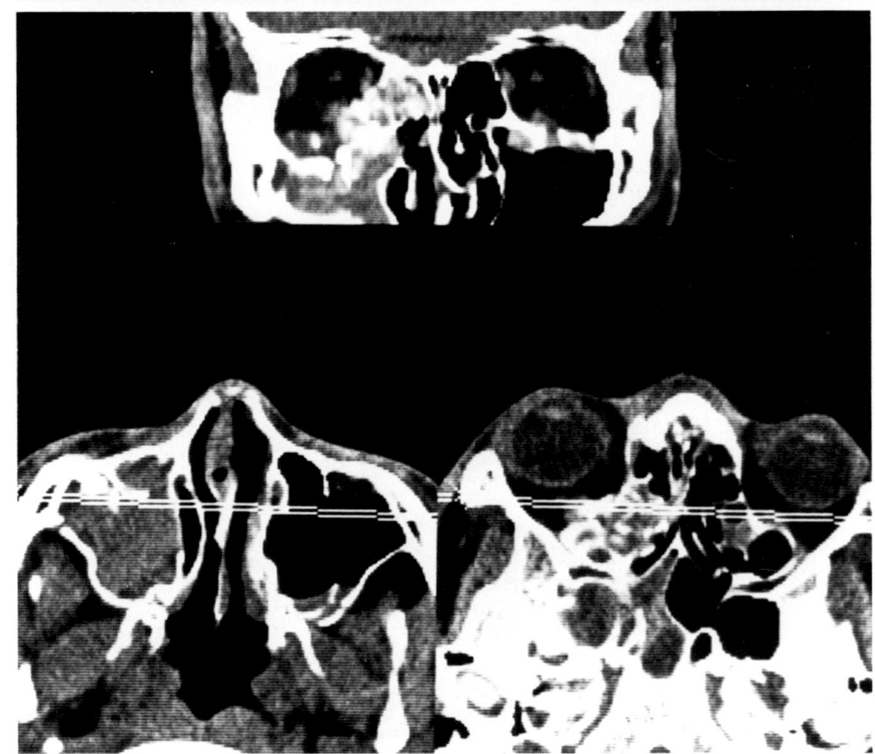

B

these data into black-and-white images that display internal anatomy on a video screen (a sonogram) (see Fig. VII–23). The image represents a slice of tissue (a tomogram), usually 1 to 2 cm thick. Because the transducer is easily positioned, the section may be in any plane (for example, transverse, coronal, sagittal, or compound). Also like CT, the image is photographed from the video screen.

Radionuclide Imaging (Scintigraphy)

Radioactive isotopes (radionuclides), such as technetium-99m or iodine-131, are attracted to particular parts of the body. Some isotopes concentrate in bone, others in specific kinds of soft tissue. The gamma rays (radioactivity) emitted from the radionuclides are recorded by a gamma camera over a collection period, typically a few seconds to several minutes, depending on the intensity of the emissions. Crystals in the gamma camera emit light according to how much radiation they receive.

FIG. VII–23: *Sonogram. (Reproduced, with permission, from Charboneau JW, Hattery RR, Ernst EC III, James EM, Williamson B Jr., Hartman GW: Spectrum of sonographic findings in 125 renal masses other than benign simple cyst. AJR 140:87–94, 1983.)*

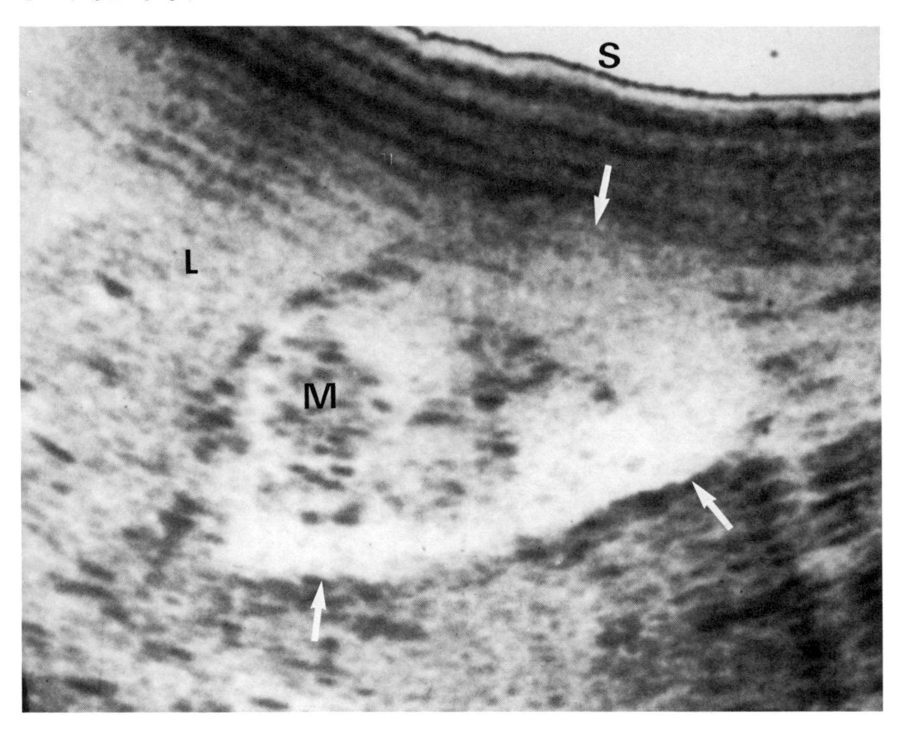

FIG. VII–24: *Scintigram. (Reproduced with permission. See figure 22A.)*

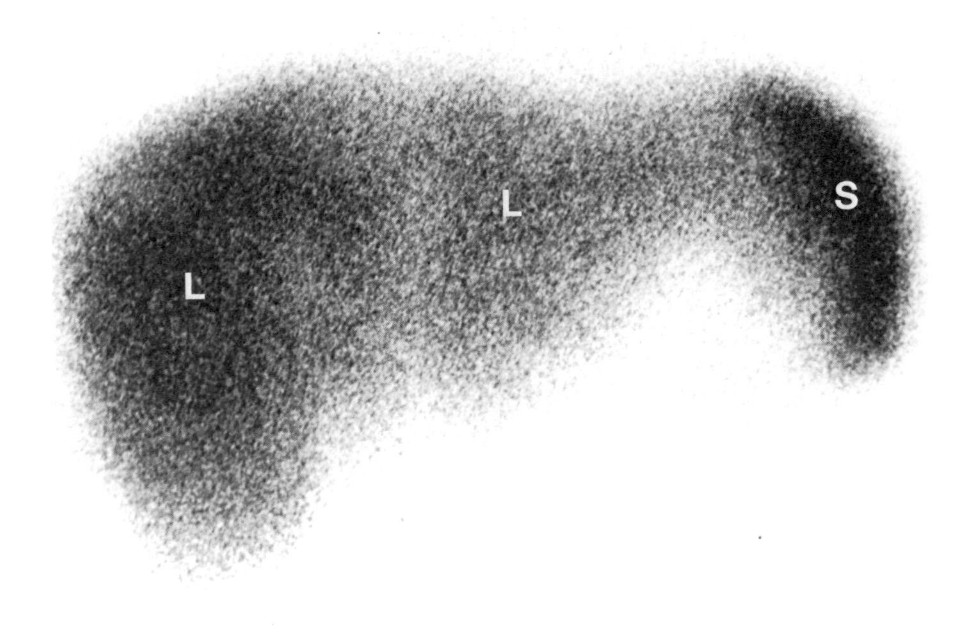

The light emission from the crystals is digitized by a computer as to intensity and spatial resolution. Subsequent translation to a black-white mode on a video screen displays the anatomic distribution of the radionuclide. Typically a scintigram shows radioactivity as black areas ("hot") and lack of it as white ("cold") (see Fig. VII–24). Like the radiograph, a scintigram is a two-dimensional display of three-dimensional information and usually is projected in the sagittal or coronal plane.

Positron Emission Tomography (PET)

Radioactive materials spontaneously emit energy in the form of alpha, beta, and gamma rays (helium nuclei, electrons, and x-ray photons, respectively). A few radionuclides also emit positrons—protons. These are characteristically emitted in pairs, which travel in exactly opposite directions from their source. If detectors are placed about an object, the positron pairs can be detected and their point of origin calculated. The assembly of points on a video screen creates an image similar to a scintigram. However, since it represents a plane of tissue, it is a tomogram.

FIG. VII–25: *Positron emission tomogram (PET). (Reproduced with permission of ME Phelps, et al., UCLA School of Medicine, Los Angeles, CA.)*

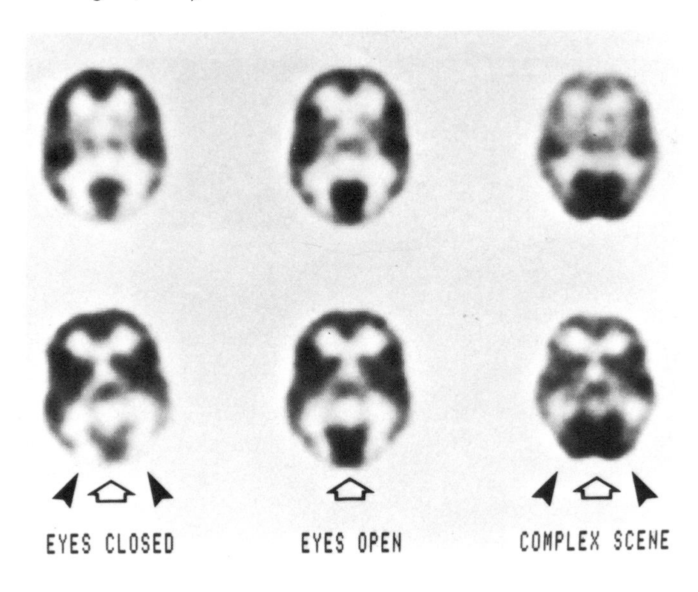

Recently, it has become possible to make radionuclide tomograms from single photon (gamma ray) emissions. The process is called single-photon emission computed tomography (SPECT). The images are like scintigrams but show two-dimensional information in any desired plane (see Fig. VII–25).

Nuclear Magnetic Resonance Imaging (NMR)

Certain atoms in our molecular structure behave as small magnets. In a strong magnetic field they are all oriented in the same axis. A pulse of energy at a specific radiofrequency will displace the atoms and cause them to spin about their axes. When the radiofrequency is turned off, the atoms return to their aligned positions, giving off the imparted energy as a radiosignal. With proper detectors, this signal can be detected and located in space, hence locating the atoms. The location of multiple atoms on a video screen produces an image very similar to computed tomography. By convention, structures with few hydrogen atoms, such as bone and moving blood, are shown in black, whereas fat and other soft tissues are white or shades of gray (see Fig. VII–26). The NMR image represents a slice of tissue (a tomogram) usually about 1.0 cm thick. It may be oriented to the transverse, sagittal, or coronal plane.

FIG. VII–26: *Nuclear magnetic resonance image (NMR). Section of brain in axial plane. (Reproduced with permission of Y. P. Huang, M.D., and P. J. Anderson, M.D.)*

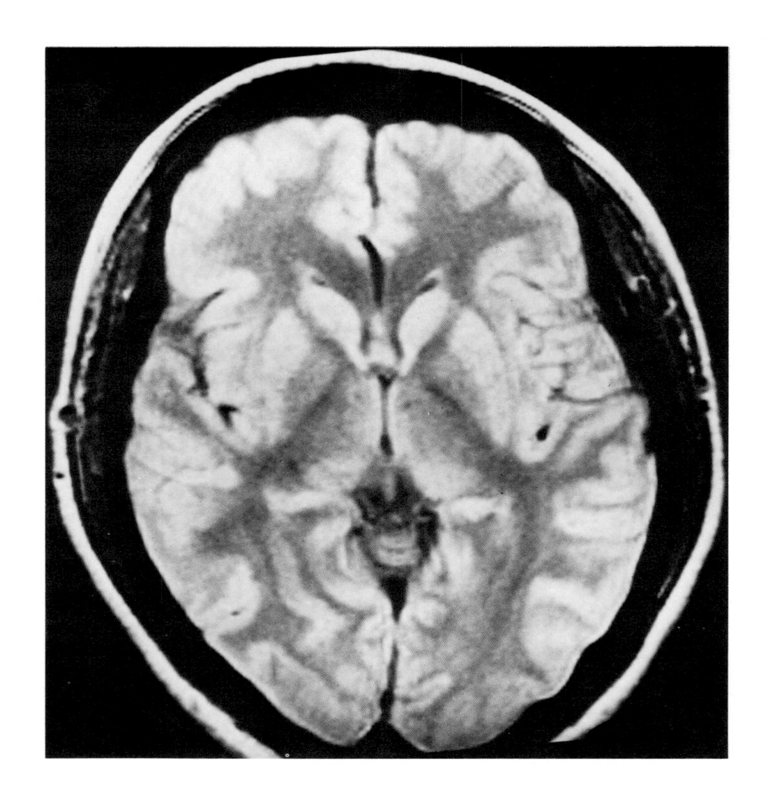

Immunofluorescence

In immunofluorescence, a fluorochrome (for example, fluorescein isothiocyanate or FITC) is attached to an antibody. A high-energy (short-wavelength) light source is then allowed to activate the antibody-fluorochrome complex. The electrons in the fluorochrome become excited and jump to a higher valence state. When the electrons return to their starting point, they give off energy in the form of light (see Fig. VII–27). This light is of a lower energy (longer wavelength) than the light that originally excited the fluorochrome. With the use of selective filters, the lower-energy light is observed in the microscope as a green fluorescence. In the laboratory, the antibody-fluorochrome complex is used to detect receptors, antigens, and/or antibodies on cells.

Karyotyping (Chromosome Identification)

In cells with true nuclei, chromosomes can usually be seen and analyzed

FIG. VII–27: *Immunofluorescent localization. (Reproduced with permission of Marshall E. Kadin M.D., University of Washington School of Medicine, Seattle, WA.)*

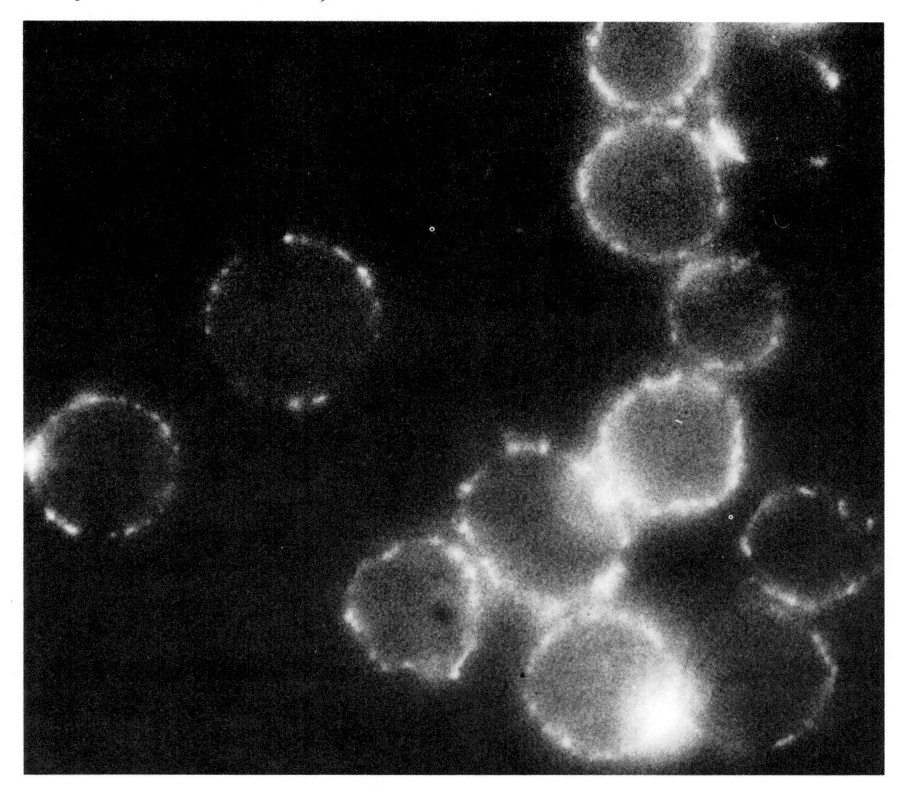

only during cell division. The chromosomes that contain the genetic material (DNA) transmitted to the next cell generation are highly condensed, threadlike organelles (measuring in the human, for example, about 7 μ in length and about 1 μ in diameter). Each chromosome in the second stage of cell division (metaphase) is formed of two identical spiral filaments (chromatids) attached by a primary constriction (centromere); during cell division, each daughter cell receives a single filament. A diploid cell possess two sets of homologous chromosomes (one paternal and one maternal).

Chromosomes can be seen after cells are arrested in metaphase with a drug (for example, colchicine). Cells are then swollen in a hypotonic solution, fixed, and spread on a microscope slide. The chromosome preparation can be stained and examined by light microscopy using a high-power lens (total magnification about × 1000). Banding techniques used for chromosome staining make possible precise identification of each chromosome. Identification is based on chromosome length, the position of the centromere, and the banding pattern along each chromosome arm. Each chromosome of a given species has a characteristic banding pattern that is formed by a succession of darkly and lightly stained areas along its length. Both light-absorbing dyes and fluorescent dyes can be used for chromosome banding.

Fig. VII–28 shows a photomicrograph of a human metaphase cell stained with R-banding. This cell from a normal male possesses 46 chromosomes, including two sex chromosomes (X and Y). The bands along the chromosomes are of various shades of gray on a white background. The chromosomes on such a picture can be cut and aligned (by homologous pairs in a diploid cell) for easier identification. Fig. VII–29 shows such a cut karyotype from a human cell stained with a fluorescent dye (Q-banding). Chromosomes stained with a fluorescent dye show white or gray bands on a black background. Rarely, a color photograph is needed to represent a banding pattern.

FIG. VII–28: *Karyotyping. Human metaphase cell stained with R-banding. (Reproduced with permission of Christine Disteche, Ph.D., University of Washington School of Medicine, Seattle, WA.)*

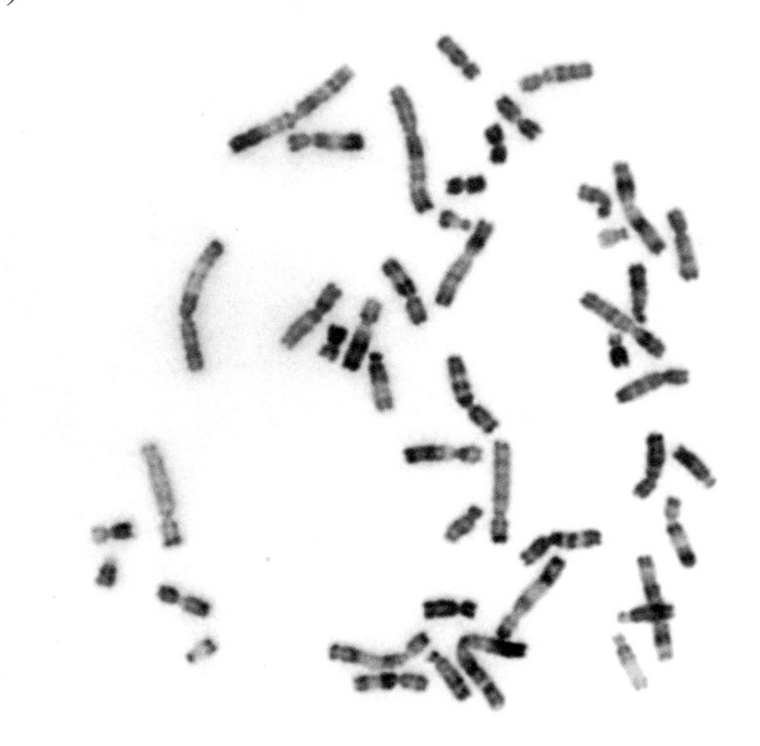

FIG. VII–29: *Human karyotype arranged to show homologous pairs of chromosomes. (Reproduced with permission of Christine Disteche, Ph.D., University of Washington School of Medicine, Seattle, WA.)*

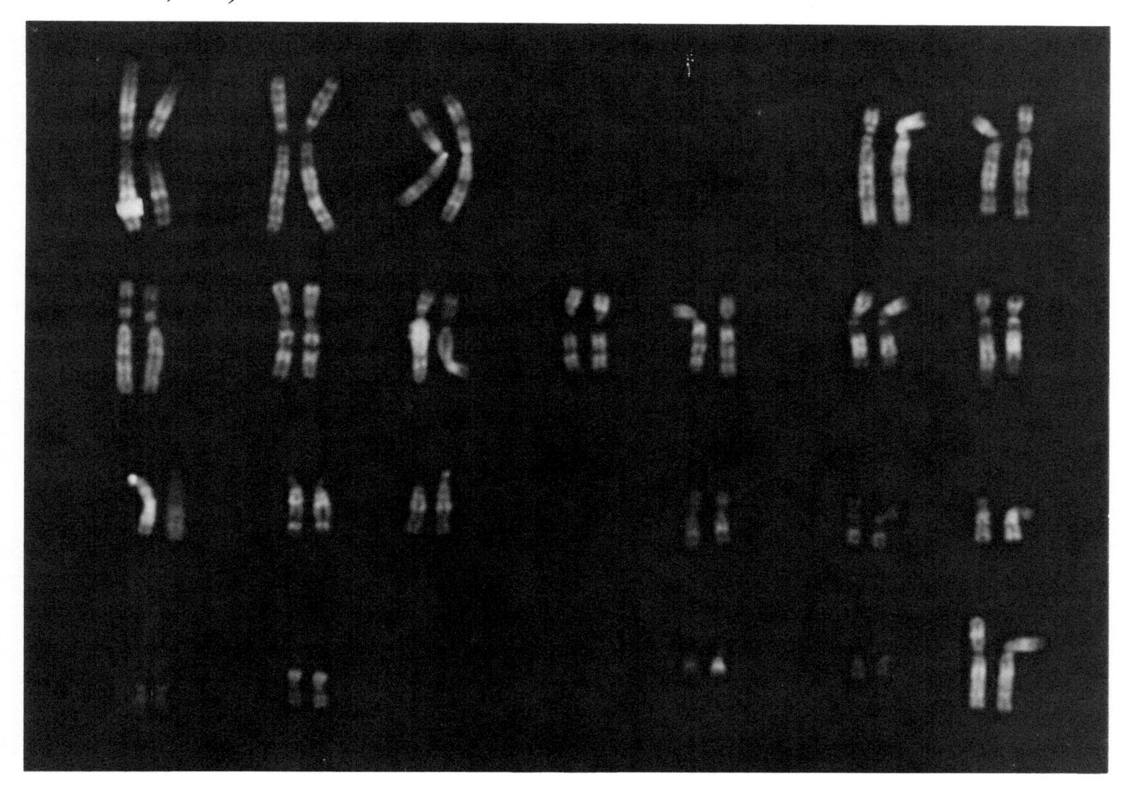

Transmission and Scanning Electron Microscopy

Electron microscopy is analogous in principle to light microscopy. Light microscopy uses a beam of light to illuminate the specimen. The light source is focused at the level of the specimen and the resulting image is magnified with a series of optical lenses for visual observation and photographing. Because the wave length of visible light is about 2000 Å, the theoretical limit of resolution (the minimal distance between two image points that can be visually discriminated) of the optical microscope is 200 μ and the maximum image magnification may vary from 1000 × to 1400 ×, depending on the optical quality of the system.

Electron microscopy uses a beam of electrons which is focused on the specimen by a series of electromagnetic lenses. Electrons diffracted from the specimen are projected on a fluorescent screen to form an image. The image is also recorded on a photographic emulsion (coated film or glass plate) which is developed as a photographic negative suitable for printing by conventional means. This type of microscopy is called Transmission Electron Microscopy (TEM). The very short wave length of an electron beam (0.5 Å in a system with an accelerating voltage of 60 kilovolts) provides a theoretical resolution limit of approximately 2 Å (0.2 μ). An image magnification range of 1000 × to 100,000 × is routinely achieved in TEM, and further photographic magnification of the electron image is commonly performed.

The higher magnifications available with electron microscopy are associated with several limitations. Penetration by the electron beam is limited by specimen thickness. Only specially prepared, extremely thin specimens (no thicker than 100 μ) are suitable. Such thin sections result in a flat, two-dimensional image. The low electron-scattering ability of most biological materials requires the use of special impregnation techniques (usually involving heavy metals) to enhance the contrast of most cellular and subcellular structures. Since only very minute areas can be visualized in a single preparation, many regions of a specimen must be sampled to ensure adequate study. The requirements of specimen preparation preclude examination of living material in such a system.

Scanning Electron Microscopy (SEM) also uses an electron beam, but in this technique the beam is "scanned" over the surface of the specimen to provoke emission of secondary electrons (similar to the action of an electron probe). Electrons are then collected by a detector and converted to a current. The latter is amplified to a signal voltage which is used to modulate the brightness of a cathode-ray tube. The cathode-ray scan is synchronized with the scanning beam in the electron microscope so that precise correspondence between contrasting light points on the cathode-ray tube and the specimen source of electrons produces an image of the specimen surface. Although only surface details are visualized, SEM accommodates much larger and thicker specimens than does TEM. The resulting images have a distinctive quality of depth and are suitable for three-dimensional reconstructions. The resolution limit of SEM ranges from 25 to 60 Å (2.5 μ to 6.0 μ). SEM magnifications are correspondingly lower than those of TEM.

Typical SEM and TEM images are illustrated in Fig. VII–30 A, B. The surface detail of a cell in tissue culture is demonstrated in Fig. VII–30 A. The finer detail of one surface projection is shown in Fig. VII–30 B.

Electrophoresis

Electrophoresis is the migration of charged particles in an electric field. When it is used as an analytical tool, small samples (typically a few microliters) of sera or solutions are applied to a supporting medium in the presence of a buffer solution. As the voltage is applied, particles with

FIG. VII–30: *Scanning and transmission electron microscopy. A. SEM preparation of mouse cell in tissue culture. The spindle-shaped cell has many processes that project from the surface of the cell. 10,000 ×. B. TEM preparation showing details of several projecting cell processes. 52,000 ×. (Reproduced with permission of Rockefeller University Press, from Robinson JM, Karnovsky MJ: Specialization of filopodial membranes at points of attachments to the substrate. J Cell Biol 87:562–568, 1980.)*

A

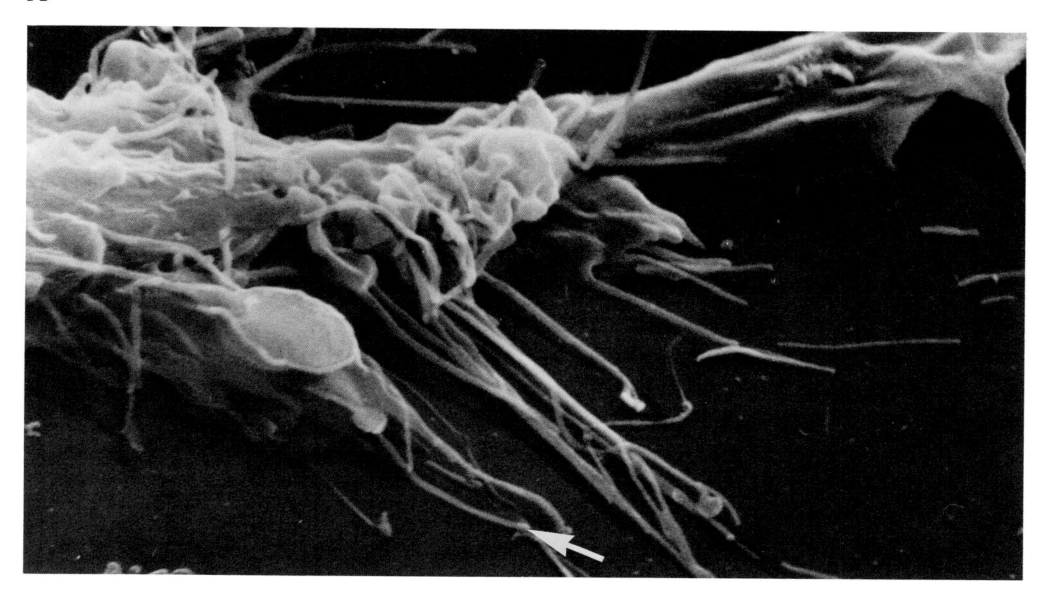

B

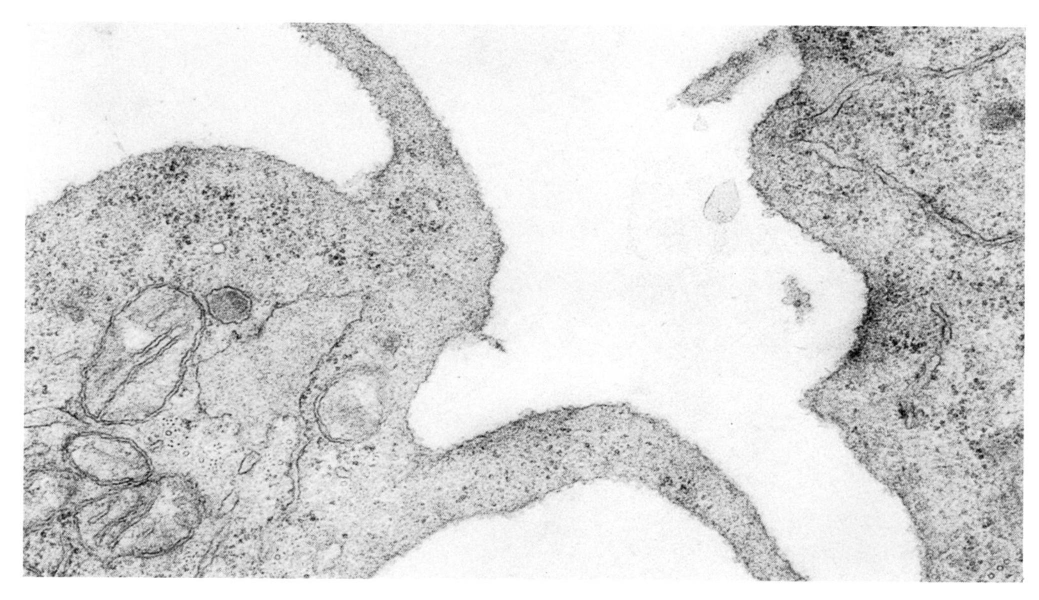

FIG. VII-31: *Densitometric quantitation of electrophoretically separated protein bands. (Reproduced with permission of Rockefeller University Press, from Cowin P, Kapprell, HP, Franke, WW: The complement of desmosomal plaque proteins in different cell types. J Cell Biol 101(4):1442–1454, 1985.)*

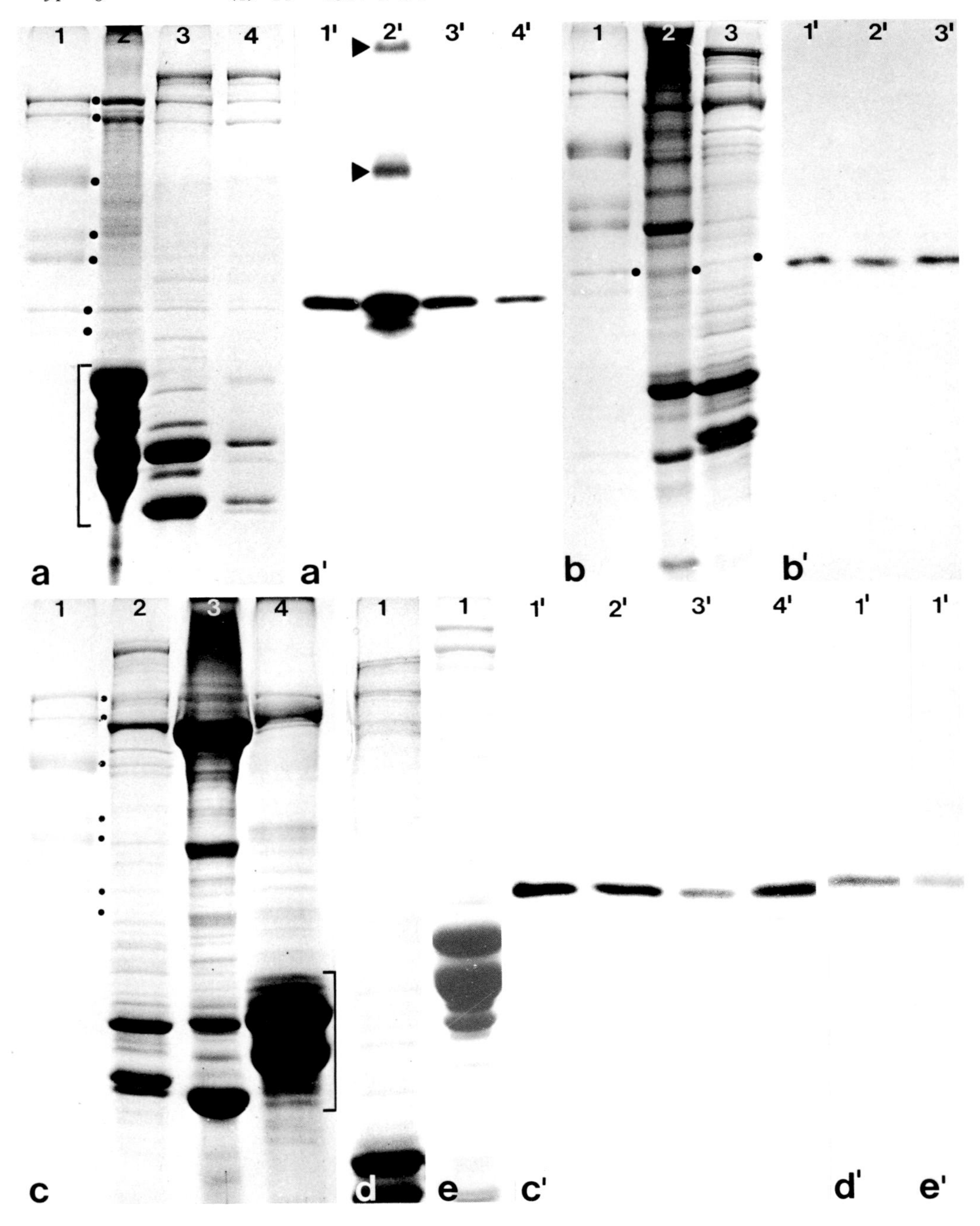

a given charge move from the point of application towards the electrode carrying the opposite charge. The rate of migration is proportional to the charge on the particle, inversely proportional to particle size, and is influenced by the shape of the particle, the nature of the support medium, and the choice of buffer solution.

In clinical laboratories, serum specimens are commonly electrophoresed on cellulose acetate membranes, and the following components are identified: albumin, alpha-1-globulins, alpha-2-globulins, beta-globulins, and gamma-globulins. Once separated, proteins in these fractions can be stained and quantitated in a densitometer, which measures the relative amount of protein in each fraction (see Fig. VII–31).

Other electrophoresis techniques are also widely used. Both immunoelectrophoresis and electrophoresis with immunofixation involve electrophoresis followed by identification of the separated fraction using antibodies against specific substances. Electrophoresis in a support medium composed of polymerized acrylamide is known as polyacrylamide gel electrophoresis (PAGE). PAGE is frequently performed on protein solutions using buffers containing the detergent sodium dodecylsulfate (SDS-PAGE) after protein denaturation. PAGE is particularly useful for assessing the molecular weight of separated substances. The technique of isoelectric focusing is a method employing special buffer solutions which create a stable pH gradient on the support medium when voltage is applied. This method separates substances on the basis of their isoelectric points, and is an extremely powerful separation technique.

Checklists of Requirements for Halftones and Photographs

Halftones

1. Halftones are used when an illustration containing shades of gray must be reproduced.
2. Finer screens result in reproduction of more detail.
3. Density of photographs can be manipulated *to some extent* to achieve best reproduction of the message.
4. Proofs of illustrations should be checked carefully, and the author must realize that the proof may be better than the printed illustration.
5. Unsatisfactory proofs can be returned for correction, but this is expensive.
6. Density and contrast of a photograph affect its reproduction.
7. There are different types of halftone reproduction (for example, duotone).

8. Printing paper strongly affects printed picture quality.
9. To do the best job, the camera operator must know what part of a picture is the most important.

Publication Requirements

1. The author must first learn the requirements of the specific publication (for example, a journal's Instructions to Authors).
2. Conventional sizes for certain types of illustrations can dictate the size of the photograph submitted for publication.
3. Individual untrimmed prints allow the greatest latitude in designing figure layout.
4. Illustrations must be identified.
5. Labels on prints should be removable to allow for changes.
6. Photographs of similar size and density can be photographed by the printer's camera operator at one time to reduce costs (gang shooting).
7. Overlays are useful to indicate area of interest, show preferred cropping, and protect the surface; overlays with labels can be shot separately, but this increases expense.
8. The convention of the specific publication should be followed in orienting a photograph.
9. Well-designed layout can enhance the scientific content of illustrations.
10. Retouching by the printer, individual print shooting, separate overlay shooting, and reshooting all add expense which must be weighed against possible improved quality.

Photographs

1. Black-and-white prints from black-and-white negatives allow best reproduction for one-color (for example, black) printing.
2. Contrast in a photograph must not be too great.
3. Photographs of CRTs should allow for sufficient depth-of-field to ensure sharp focus.
4. Color slides should be converted to panchromatic transparencies for black and white reproduction.
5. Over-exposure, under-development, and electronic scanning can be used to obtain an optimal photograph of a diagnostic image (for example, a radiograph).
6. A photograph should be about one fifth larger than its final printed size (reproduction of about 80%).

7. Type of photographic paper can affect the quality and reproducibility of a photograph.
8. Photographs can be retouched to eliminate flaws or dust, but scientific content must not be changed.

Primary Images

1. Different imaging procedures result in different types of primary image.
2. Radiography produces a direct transparency.
3. Other methods, such as CT and NMR, produce images from digitized data on CRT screens.

CHAPTER VIII
COLOR ILLUSTRATIONS

The Use of Color

In recent years there has been a dramatic increase in the use of color in publications. Both advertising and editorial color have increased rapidly. It has been suggested that this increased use of color has come about because of a subtle demand by a new generation of readers, advertisers, and publishers. This new generation has grown up with color television and is conditioned to colorful images, whereas those of earlier generations were satisfied with black-and-white reproduction. Technology and increasingly attractive economics have also increased the feasibility and cost-effectiveness of color reproduction. Whatever the cause and effect, the use of color in illustrations is commonplace today. Because there is every indication that this will be a continuing trend, it is incumbent on the publishing community to develop a working knowledge of color reproduction.

The Color Process

In printing production there are two techniques for reproducing color: spot color and the four-color process.

Spot Color

Spot color is the use of a colored ink to add color to an illustration. It is generally used in conjunction with a black-and-white illustration, but can also be used independently, depending on the desired effect. Most commonly, spot color is effective in graphs, charts, and maps (see Chapter IV, Maps, Color, page 129). Enhanced effects can be realized with the use of multiple spot colors for delineating quantitative values. The number of spot colors available is dependent upon the configuration of the printing press. Spot color can be used to highlight a particular segment or segments of an illustration (for example, veins and arteries). The eye is drawn immediately to the subject at hand (see Figs. VIII-1, 2; Fig. III-44).

An additional use of spot color is to enhance the appearance of a black-and-white halftone. Combining one other color with black

FIG. VIII–1: *Use of single spot color to highlight overlapping distributions in an area map. The key identifies the distributions of species of the Castilleja viscidula group (modified from Fig.* IV–31, *page* 126.)

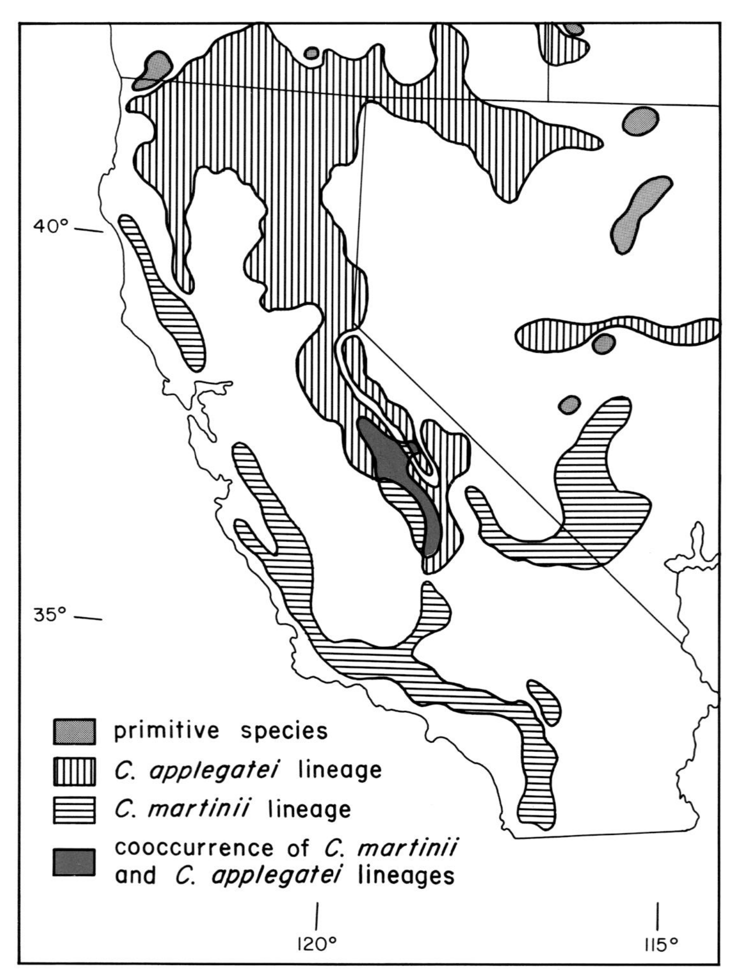

through a special screening process results in a two-color halftone, known more commonly as a "duotone." Particularly in the medical field, black-on-black duotones are used to enchance depth and richness by increasing contrast while maintaining detail (see Fig. VII–20).

To create a spot color illustration a separate piece of mechanical art must be prepared for each different color. This mechanical art is used by the printer as camera-ready material.

A wide range of spot colors is available to the scientific illustrator. Printing inks can be mixed to match the required hues and tones. The standard reference in the industry is the PMS System (Pantone® Match-

FIG. VIII-2: *Use of single spot color to emphasize form in a botanical line drawing. Habit of Cavendishia pedicellata (Luteyn et al., 7308). Colombia, Choco, Ansermanuevo-San Jose del Palmar rd., 2–5 km E of San Jose del Palmar, 1200–1500 m, 20 APR 1979. (Reproduced, with permission, from Luteyn JL. Ericaceae Part I. Cavendishia. Flora Neotropica Monograph 35. New York, The New York Botanical Garden, 1983.)*

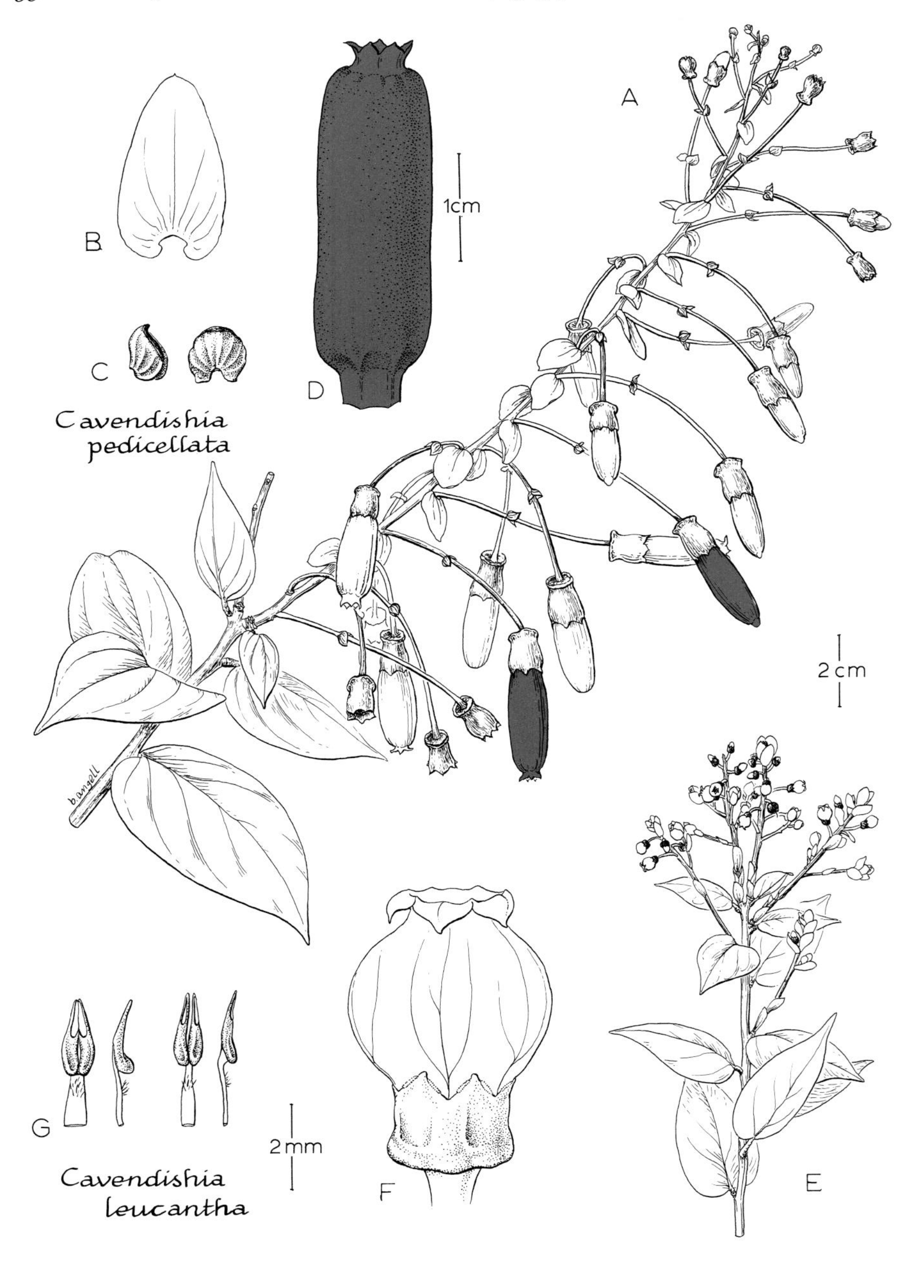

ing System). A PMS book contains swatches of nearly 600 colors on both coated and uncoated paper, each with their ink-mixing formulas utilizing eight basic PMS ink hues. The color samples simulate the end result but are not necessarily an exact duplicate of the original. The book is available from the publisher (see Literature Cited) and at art supply stores.

Strong colors that allow various percentages of screening or hues are preferable as spot colors. Red, for example, is a good spot color; yellow is not. Spot color is usually the least expensive method of adding color to an illustration, when limited to one or two colors, as less preparatory and press work is required than for process color (see below). Spot color is also available through the four-color process, although it can be more costly.

Four Color

The four-color process, more commonly known as "process color," brings the full spectrum to color reproduction. Process color also brings with it its own set of complexities and economics, which must be considered in decisions concerning its use in scientific illustrations. Its most common use is in reproduction of color photographs (either transparencies or prints), drawings, and paintings. Some illustrations demand process color; for example, the subtleties of stained smears in microbiology can be accurately represented only by the four-color process. The medical illustrator may also use process color in graphic representations of anatomic material, and certain pathological conditions sometimes require four-color illustrations for diagnostic purposes. It should be noted that color slides and reflective color artwork lose impact and quality when reproduced as black and white illustrations. Color tone gradations lose their subtleties, and color differentiation suffers, when converted to black and white.

There are times when the complexity of the subject matter to be illustrated may dictate that four-color reproduction is inappropriate. An alternative in these instances is to reproduce the illustration in black and white, identifying by numbers the detail to be referenced.

So that material can be properly prepared for process color reproduction, it is important to become acquainted with the separation process. Process color is the breakdown or "separation" of a multicolor image into four process colors: yellow, magenta (a red violet), cyan (a peacock blue), and black. When these images are combined on the press, the original subject is reproduced with remarkable fidelity. The variety of colors possible through combinations of various intensities of the four process inks is almost unlimited. (see Fig. VIII-4).

Within the separation process it is possible to modify or "color-

FIG. VIII–3: *Four color build up on web press. (Reproduced, with permission, from Impositions for Web Offset Publications. Richmond, VA, William Byrd Press, 1983.)*

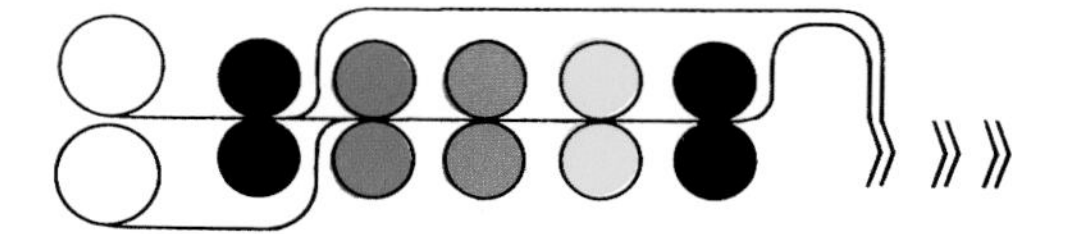

5 UNITS, 2 WEBS—24 and 32 pages
1 color 2 sides top web and
up to 4 colors 2 sides bottom web

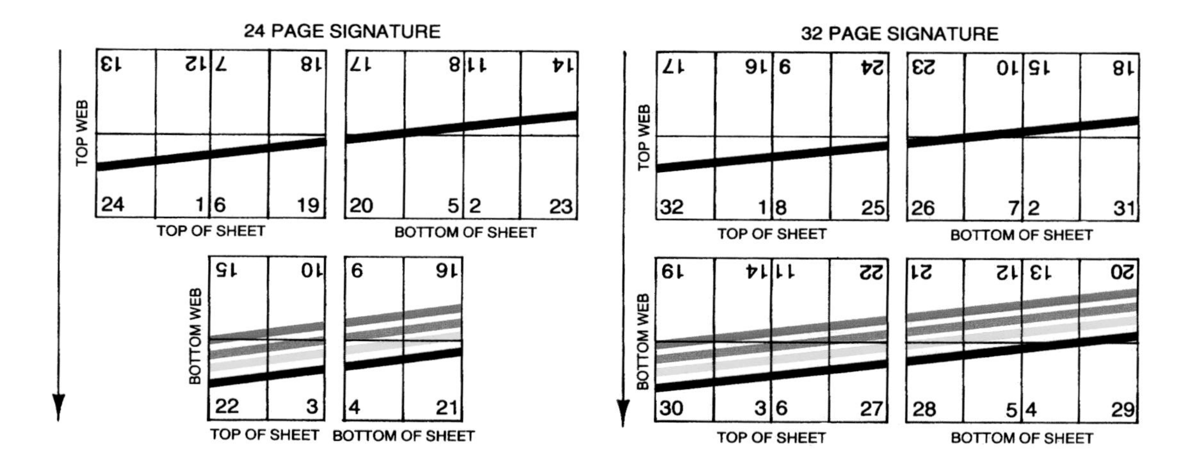

correct" hues. The author or editor should consult with the separator or printer to determine in what form and with what terminology instructions for modification should be conveyed. The original material must then be returned for the correction. Electronically or mechanically, depending on the separation system used, the individual separation negatives will be modified. Some printers have in-house separating capabilities whereas others use outside resources.

Since color is largely subjective, it is usually described as seen through the eye of the beholder. Generalized instructions from the author or editor to the printer ("beef up the color"; "color too skinny, make it heavier"; "color too strong, thin it down"; "reduce overall"; "increase detail") will not effectively communicate the job requirements. Precise instructions can be given with a color matching system, such as the Selectone® Four Color Process Guide or the Pantone® Matching System, to specify the colors needed for reproduction ("print background as 80% blue, 20% red"; "match foreground to 3% Pantone® yellow, 97% Pantone® white"). Specific instructions written on a transparent overlay, identifying the critical areas and the primary colors to be emphasized, are especially helpful to the printer.

The separation process is accomplished by both mechanical and electronic scanning processes. Electronic scanning is rapidly replacing mechanical processes. In scanning, the artwork is processed by electronic or mechanical filter of other basic colors to produce a single negative for each of the four colors described above. In the four-color printing process, each negative is used to produce a single printing plate for each of the respective colors. The printing process requires a separate press run for each color on a single-color press or a separate printing unit for each color on a multi-unit press.

In process color there are two general types of copy: reflection and transmission. *Reflection copy* includes color photographic prints, color drawings and paintings and graphics such as diagrams, graphs, and charts. *Transmission copy* refers to color photographic transparencies, such as 35-mm slides.

Color quality considerations play the critical role in the entire process, from the characteristics of the original copy through its final reproduction in printed form. The final product can be no better than the original material. The printer should receive only original material that is satisfactory to the author. It is usually easier to evaluate the quality characteristics and reproduction potential of reflection copy, since the production will be viewed the same way—with reflected light. Many publishers prefer reflection copy because of this commonality and the fact that everyone is seeing the same medium at the initial stage. Reflective copy is also easier to work with when applying labels, scale markers, lettering, and so forth. Publishers view these as offsetting advantages to the disadvantage of second-generation degradation of transparencies converted to reflective copy.

Color transparencies are more difficult to evaluate because of their wider contrast range, and greater color intensity, and the unstandardized conditions for viewing transparencies by transmitted light. They should be examined in a viewing environment that enables accurate analysis of the characteristics required for reproduction. This examination should never be done in a darkened room. The ideal method is to use a viewing booth in which the viewer can look at a projected (and therefore reflective) image.

The standard transparency illuminator for viewing 4 × 5 inch transparencies or larger provides a surface luminance of 1400 ± 300 CD. This is equivalent to 409 ± 88 foot Lamberts as measured with either a luminant meter measuring an area having a diameter equal to 1/20th of the shortest linear dimension of the illuminated surface area or with a luminant meter in contact with the front surface of the transparency. The illuminator surface should provide diffused light such that the luminance of the surface measured at any angle between 0 and 45 degrees from normal shall not be less than 75% of the luminance of the same area as measured normal to the surface. The luminance should ideally be the same at all points on the surface. Any departures from complete uniformity must be even or gradually diminishing from center to edge such that the luminance (measured at a normal angle) at any point within the illuminated areas is not less than 75% of the luminance measured at the center of the viewer.

Whether reflective copy or transmission copy is to be examined, certain quality considerations should be addressed. Scientific illustrations are often the most difficult to reproduce accurately, and more is expected of their reproduction; the original copy must therefore be of the highest quality. The original material should be sharp and clear, as the printing process cannot compensate for the absence of these attributes. The material should be neither over- nor underexposed, and highlights and shadows should contain proper detail. Color should be balanced to provide a pleasing effect. If multiple illustrations are to be used, they should all be consistent in quality. If complete consistency is required, the same type of film and even the same emulsion batch of that film should be used to record the primary image. Enlargement factors should be similarly matched to prevent variation in graininess.

Despite the viewing disadvantage, transparencies are often used for reproduction. They are less expensive than color prints; however, they do not give a precise representation of what will be reproduced. Without having seen the original artwork it is difficult to project accurately what the colors should be in the finished product. Additional printer's expense may also be involved in applying labels and internal markers for scientific subjects. Since the original color image is often submitted as a 35-mm transparency, such labeling cannot usually be done by the illustrator, author, or editor. Positive color prints, on the other side, are easily labeled before color separations are performed.

There are few restrictions on the sizes of transparencies submitted for separation, either by photographic technique or by electronic scanning. This is also true for photographic separation of reflection copy but not for scanning, for which size and flexibility requirements are more restrictive. However, some electronic scanners can accept original copy as large as 30 to 40 inches. Reflection copy must be flexible enough to permit wrapping around a cylindrical drum. Any copy that is extra large or small, thick or rigidly backed, or oddly shaped, or that requires extreme enlargement or reduction, calls for special handling and therefore incurs extra cost. Mechanical separations have been made from copy as large as 4 by 8 feet. Wherever possible, material of similar characteristics should be color separated at the same time, to take advantage of economic ganging opportunities (grouping for simultaneous processing). It is a good idea to consult with the printer or separating facility when planning the illustrations, to determine any material restrictions.

Extra caution should be taken in handling all original copy. It is difficult, if not impossible (and always costly), to repair copy damaged by mishandling (see Chapter IX).

Proofing

Proofs of color reproductions are a critical step in the reproduction cycle. They are the primary means of error and color deficiency detection, they identify potential reproduction problems (and suggest solutions), and, in their final approved form, they serve as a clear statement of what the publisher expects. As noted above, both spot color and process color separations can be color-corrected or altered, by either photographic dry dot etching or a chemical process called dot etching, or by computer image enhancement and alteration in the scanning process. Cost considerations dictate the degree to which these capabilities can be exercised.

All proofs should contain color bars incorporating a repeating pattern of solids of single inks and two- and three-color overprints. These color bars are later compared to printed material on press to verify color ink densities (see Fig. VIII–4).

There are various methods of proofing which differ in effectiveness, desirability, and cost. The most desirable are press proofs. When produced on a similar press with the same inks and printed on paper stock identical in color and surface characteristics to the stock that will be used for the publications, press proofs will yield the closest representation of what the final printed product will look like. An even more accurate result can be obtained by using the same type of plates that will be used in the production run, printed on the same or a similar press and in the same wet or dry-on-dry ink modes. Press proofing is by far the most expensive proofing method; however, no other method can more accurately replicate actual press production.

From a more economical and practical point of view there is no better way to achieve desired results (or lose them) than at the litho preparatory phase of the printing process. A proof at the pre-press stage strives to illustrate, within inherent constraints of proofing, the final product. Furthermore, it pinpoints problem areas that must be addressed before going on press.

A more popular and effective litho preparatory photographic proofing method is the Cromalin® process. This process involves lamination of separate layers of color-impregnated film into a one-piece proof.

The Matchprint System® combines the dot-for-dot consistency with a new coating and adhesive technology. It consists of four individual color proof layers which can be applied directly onto the paper stock specified for the publication. It also simulates dot gain that might result from printing.

Color key and match key proofs use separate sheets of film for each of the four processes. These can be viewed over a sample of the publication paper stock or a close substitute, giving an approximation of what can be expected on the production press run. Both of these off-press methods can be used for spot color proofs; however, the capability of matching inks is fairly limited.

When a desired proofing method is not available, four-color process subjects can be printed on a single- or a two-color press. Under these conditions, the printer usually requests that the color separator supply progressive proofs. After the separations have been completed, the separator makes printing plates, one for each color (four in all), and press proofs each of these in sequence. The resulting progressive proofs show the addition of each color as it is laid down on the sheet to complete the four-color image (see Fig. VIII–4). Thus, progressive proofs are a set of four sheets of paper showing the printing of the yellow; yellow and magenta; yellow, magenta, and cyan; and yellow, magenta, cyan, and black separations. This proofing process enables the printer to check each color pass on the press as he lays that ink down, and to make whatever ink adjustments are necessary. Careful control of these intermediate printing steps will ensure that the completed printed product matches the fourth progressive proof.

A good practice is for the printer to supply the editor or publisher with the fourth (final) progressive from the separator, to serve as authors' proofs. The color proof, in whatever form chosen, is the critical vehicle for the author and editor to use in communicating requirements to the printer. The printer uses the proof at the press as a guide to achieve the desired results. It is often necessary and advantageous to go through multiple separating, color-correcting, and proofing cycles. This can be costly, however, and economics usually dictate the degree to which it is desirable or feasible.

It is customary for the printer to return the original artwork to the publisher as soon as practical after the printing process has been completed.

Industry Color Printing Standards

There are printing trade standards for production of process color, reproductive film, and proofs, designed to achieve the best possible reproduction throughout a publication. Each printer may have certain copy requirements and quality procedures of which author and editor should be aware. The most comprehensive standards published were developed by a consortium of publishing, advertising, and printing industry organizations; they cover film preparation requirements and proofing requirements. These standards are the SWOP standards (Recommended Specifications Web Offset Publications), and are available through the organizations referenced below. They are updated and republished periodically.

FIG. VIII–4: *Progressive proofs showing the cumulative effect of adding color separations. A. Yellow. B. Yellow + magenta. C. Yellow + magenta + cyan. D. Yellow + magenta + cyan + black. Note the color test strip which verifies color density and registration. (Reproduced, with permission of Storey Communications, Inc., from Proulx A, Nichols L, Sweet and Hard Cider, 1980. Compact color test strip reproduced with permission of Graphic Arts Technical Foundation.)*

A

B

C

D

GATF

YELLOW MAGENTA CYAN BLACK GATF COMPACT COLOR TEST STRIP

FIG. VIII–5: *Out-of-register printing of Fig. VIII–4 D. This may result from failure to maintain accurate registration of the overlaid color screens.*

Other Quality Considerations

Certain design characteristics should be avoided. Small type, especially with fine serifs, printed in more than one color or reversed out of a multicolor background are subject to mis-registration and may therefore be difficult to read. It is best to produce all reverse lettering with a minimum of color. No less than a 70% screen should be used for the shape of letters, and letters to be printed in nondominant colors should be printed slightly enlarged, to help minimize the problems associated with registration. Avoid heavy solid areas which may cause uneven inking resulting in streaks and a phenomenon called "ghosting" (ink starvation on a page).

In selecting the paper stock for a publication containing color illustrations, it is important to remember that coated stock reproduces the best color fidelity. The use of uncoated stock results in significant degradation of color reproduction.

The imposition or placement of color pages in a printing form is critical to successful color reproduction. Color pages must be placed so as not to conflict or compete with other color pages for ink. Color conflicts occur when two color pages, in line with each other, have quite disparate ink requirements in one or more colors. In such cases the printer must "favor" the color on one illustration at the expense of the same color on the other illustration. For example, printing the proper amount of red to produce an acceptable flesh tone in one illustration may result in a solid red-appearing pink in the other illustration. Effective color proofing and adherence to industry standards helps prevent or reduce in-line problems.

Caution must be exercised when an illustration crosses over two facing pages. Because of inherent folding mechanics in some press operations, two different forms could each be folded more or less than 1/16th of an inch and the resulting illustration could be misaligned in the cross-over by as much as 1/8th of an inch. This problem does not occur when the cross-over illustration is confined to facing pages within the same signature.

There is a phenomenon in the printing process called dot gain, in which the size of the printed dot on the separation or negative increases through the plate-making and printing processes (see Chapter VII, Dot Gain, page 197). A dot can gain as much as 10%, resulting in a much darker reproduction than intended unless corrective steps are taken. When making negatives and separations, the printer takes dot gain into consideration. However, during the production run the dot gain may not always be consistent throughout the run or throughout the illustration. This affects the decision as to the type of equipment on which a critical scientific illustration should be reproduced. Considerations of dot gain should also influence the choice of paper, since uncoated stock generally produces greater dot gain than coated stock.

Although the printer usually goes through a series of inspection and control steps throughout the entire process, including the pre-press proofing techniques and on-press control techniques, the publisher may want to reinforce the quality control procedure by examining sample signatures before they are bound into the final publication. On-press inspection can be expensive, since the printer may charge the publisher for idle press time while corrections are being made. Again, this is a question of economics.

Occasionally a coating of varnish is applied to an illustration, particularly on cover illustrations, to provide an extra measure of protection to the quality and life of the illustration. Varnish can be applied as a part of the normal printing process, given the availability of an extra printing unit on the press.

Although more expensive, metallic inks, such as silver, can be used for special effects. However, these are not recommended for halftone reproduction or fine reverses.

Certain other manufacturing considerations should be recognized at the time of designing and specifying color illustrations. This chapter has addressed the use of spot color and process color as though they are mutually exclusive processes. Although this is true, they can nonetheless be combined. It is possible to augment a four-color process illustration with a spot color or colors. This is usually done when a particular color for a specific portion of the illustration is not reproducible from the original copy by the four-color process.

Economics and Planning

From an economic standpoint the publisher should attempt to confine color illustrations into the same form or forms. Each time a black-and-white form is opened up to color, production costs increase significantly. The key is to consolidate or "piggyback" color on forms by designing the pagination so that colors fall on the same forms on the same side of the press sheet. Some scientific publications cluster articles with four-color illustrations in signatures that contain four-color advertising. If this is possible, the cost advantages are significant (see Fig. VIII–6).

The arrangement and complexity of spot color can heavily influence the litho preparatory cost of the publication. It is best to seek advice from the printer before proceeding. Since the color separation process is an additional production step, time should be allowed for this in the publishing schedule.

Another way of economizing is to take advantage of a printing feature called split fountains. The ink fountain of some presses can be divided or split into two or more sections, each section holding a different color of ink. This allows the printing of multiple colors from a single press cylinder, at a fraction of the cost of using multiple cylinders.

FIG. VIII–6: *Planning impositions to reduce costs. The diagram shows a 16 page 4 color signature on a 4-color unit web press. By placing all of the color pages on one side of the sheet the cost of set up for 3 additional units is saved. (Reproduced, with permission, from Impositions for Web Offset Publications. Richmond, VA, William Byrd Press, 1983.)*

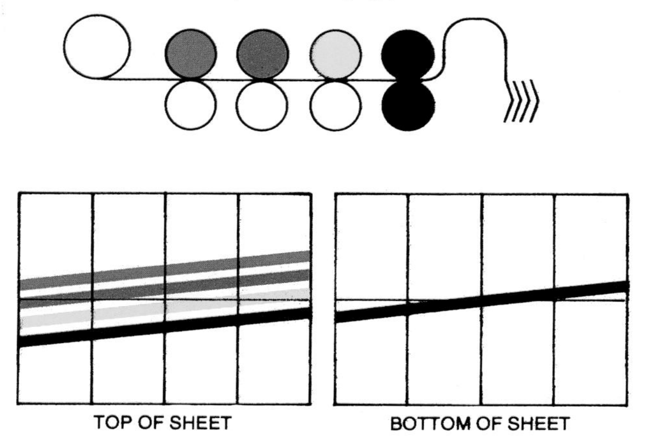

FIG. VIII–7: *Example of a production schedule. Production planning is facilitated by specifying deadlines for each stage of production.*

Journal of ABC — May 1988 #__________

ABC = Publisher
XYZ = Printer

Description	May	Notes	Working Days
1.1 Manuscript at XYZ	2/4	25% (350)	
1.2	2/12	25% (350)	1.1→1.2 = **7**
1.3	2/19	20% (280)	1.2→1.3 = **6**
1.4	2/26	20% (280)	1.3→1.4 = **6**
1.5	3/9	5% (70); **Cut-off date for color subjects: 2/9**	1.4→1.5 = **9**
2.1 Pages leave XYZ	2/23	Unfolioed pages	1.1→2.1 = **14**
2.2	3/2		1.2→2.2 = **14**
2.3	3/9		1.3→2.3 = **14**
2.4	3/14		1.4→2.4 = **12**
2.5	3/17		1.5→2.5 = **7**
3.1 Pages at ABC	2/24		2.1→3.1 = **2**
3.2	3/3		2.2→3.2 = **2**
3.3	3/10		2.3→3.3 = **2**
3.4	3/15		2.4→3.4 = **2**
3.5	3/18		2.5→3.5 = **2**
4.1 Pages at XYZ	3/8		3.1→4.1 = **10**
4.2	3/16		3.2→4.2 = **10**
4.3	3/21		3.3→4.3 = **8**
4.4	3/23		3.4→4.4 = **7**
4.5	3/25		3.5→4.5 = **6**
COVER black and white:			
5.1 at XYZ	3/9		same as 1.5 date
5.2 stat leaves XYZ	3/16	stat only to AU and ED, include cover legend	5.1→5.2 = **6**
5.3 stat at XYZ	3/25	See Line 8	5.2→5.3 = **8**
COVER color:			
5.1 at XYZ	2/9	See note 1.5—order seps—two chromalins	color cut-off date
5.2 chromalin leaves XYZ	2/22	to AU and ED	5.1→5.2 = **10**
5.3 chromalin at XYZ	3/4	See Line 8	5.2→5.3 = **10**
6 Final Material at XYZ	3/23	Mini Commentaries 2½% of total ms. (35)	same as 4.4 date
Contents		Order of Articles—Including page count for item 10	
7 Extra material at XYZ	3/31	± 10 printed pages 2½% of total ms. (35)	4.5→7 = **5**
		Page makeup previously known—Mini Rev., Mini Comment., etc.	
8 Folioed "revised" pages leave XYZ	3/31	Includes Final Material, Contents, Front Cover	4.5→8 = **5**
			6→8 = **7**
9 Folioed "revised" pages at ABC	4/1		8→9 = **2**
10 Index at XYZ	/	On disk—June and Dec. issues only	9→10 = **3**
11 Folioed "revised" pages at XYZ	4/6		9→11 = **4**
12 Extra page proofs leave XYZ	4/6	Final folioed pages. Includes index (no revise)	7→12 = **5**
Phone OK	4/7		Index issues: 7→12 = **7**
13 Print order/offprint order at XYZ	4/7	Press Release Date	11→13 = **2**
14 Extra proofs at XYZ	4/8	Confirming	12→14 = **3**
15 Mailing Labels at XYZ	4/21	3 days before Mail date	13→15 = **11**
To Planners	/		
16 Mail date	**4/26**		13→16 = **14**

- Working Days are inclusive. There are **59** Working Days in this schedule.
- XYZ Holidays include Memorial Day, July 4, Labor Day, Thanksgiving (2 days), Christmas (2 days), and New Year's (2 days). Also include Federal Holidays not listed as XYZ Holidays.
  Please submit manuscript early according to the number of holidays falling within this schedule.

Of major importance to the achievement of desired results in color reproduction in scientific illustrations is the degree to which good planning and good communication are accomplished by all parties involved. Good planning starts with a planning meeting at which the publisher, editor, and printer communicate firsthand the expectations and constraints for a given publication and process (see Fig. VIII–7). These understandings should be well documented and all specifications clearly detailed. In conjunction with these discussions, dummies and imposition sheets should be available for planning layouts. Cost estimates should be reviewed and, where appropriate, economic constraints that affect illustrations should be conveyed to the publisher. Proofing techniques should be discussed and agreed on, as well as production schedules that allow sufficient time to accomplish the desired quality assurance steps. The key to a successful publication is totally open communication throughout the entire planning and production process. Expectations, understandings, and instructions must be clearly articulated and set forth in writing. Costs should be well understood and similarly documented.

LITERATURE CITED

American Association of Advertising Agencies: Recommended Specifications for Web Offset Publications. American Business Press, Magazine Association, 4th Issue, February 1981

Byrd Press: Impositions for Web Offset Publications, Products and Services Manual. Richmond, VA, Byrd Press, 1983

Byrd Press: Planning Your Publication — Impositions. Richmond, VA, Byrd Press, 1972

International Paper Company: Pocket Pal: A Graphic Arts Production Handbook. New York, International Paper Co., 13 ed., 1983

Pantone, Inc.: Pantone Color Formula Guide. Moonachie, NJ, Pantone, Inc., 18 ed., 1985

CHAPTER IX
MATERIALS AND HANDLING

Ground Materials

Drawing surfaces are described as ground materials and they include a wide variety of papers and films that change regularly as a result of new manufacturing processes, shortages, and the development of new techniques. The illustrator should select ground materials that will best accommodate the technique to be used. Some ground materials are textured; some are smooth. There are thick and thin surfaces, stiffly premounted or flimsily free, coated and shiny, or transparent. Most lightweight ground materials should be securely mounted on a heavier backing, and a cover sheet should be used under the artist's hand to avoid smearing or damaging the surface while the drawing is being rendered.

A textured surface will not appear rough when photographed, but a line drawn on it will probably reproduce as a wiggly line. Drawings on translucent materials, such as plastic film taped to an opaque backing, sometimes cast shadows onto the mounting board. However, if the drawing is photographed under glass and/or in a vacuum frame with balanced lighting, this shadow will not appear in the final print. When the artwork is to be viewed directly in an exhibition, the illustrator may apply white paint to the back of the film to eliminate the shadows, but this should not be necessary for reproduction.

Permanence

Longevity of ground materials should be a consideration in preparing all artwork, since some manuscripts are held for a year or more before being published and some artwork is of archival importance. The more acid there is in a ground material, the shorter its life will be. It will fade, turn yellow, become brittle, and/or disintegrate after several years. Papers with a low or nonexistent content of rag or cotton may be acid free, whereas papers made from wood pulp may have high residual acid (unless it has been specially treated, which is unlikely with most inexpensive art materials). Other ingredients, such as sizings, bleach, etc., further contribute to paper's deterioration. High rag content papers and boards, or those labeled "acid free" or "pH neutral," last the longest of all. Unfortunately, some illustration boards labeled "100% rag" have an acidic core and backing which will eventually damage the drawing surface. If the manufacturer has not specified the rag content and pH of the paper, it is probably of low quality and will deteriorate rapidly.

The environment also can increase acidity in paper: air pollution, cheap wrappers, fingerprints, rubber cement, and tapes other than those sold for archival uses. Heat, humidity and darkness encourage mold. Boards coated or embedded with calcium-containing materials, such as clay, are fairly durable, and the thicker the coating the better.

Polyester films are made of polyester resin and seem to be completely inert and durable. Acetates age because acids are used in the manufacturing process, and they also contain plasticizers which eventually evaporate, leaving behind a brittle surface. Some illustrations prepared on photographic papers and with various chemicals may fade within a year or two if exposed to light.

All drawings and paintings should be protected from dust and dirt by a cover sheet. The texture of paintings is especially prone to catch dust. Artwork should always be handled only by the edges. The surface of any drawing should never be touched. Fingerprints leave reproducible marks on some surfaces, and certain drawing media smudge easily. Paper clips and other attachments must be kept off of drawings, including the margins. Paper clips can leave an imprint after only a few minutes, and may eventually rust onto the surface. A loose paper clip that slides over the drawing surface can mar the drawing.

Scratch Board

Scratch board is a coated board commonly used for line artwork. It is also called scraper board. It consists of several plies of paper bonded together and spread with a thin, smooth coating of white clay mixed with binding agents. The illustrator can paint an area with India ink, or can use black scratch board that has an extra coating of ink, and draw or etch on it with a sharp knife blade. Corrections are made by scratching the unwanted ink off the surface with the knife. Scratch board cracks easily, so it must be mounted on a stiff backing before use. Protection of the surface in the studio can be provided with a flat piece of cardboard laid on top of a covering sheet. When the drawing is shipped or mailed it should be protected on both sides by heavy cardboard.

Clay-coated Stipple Boards

Clay- or calcium-coated boards other than scratch board may have a textured surface that will hold pencil dust. The texture is often very fine and practically invisible to the naked eye. Carbon and graphite dust drawings on these surfaces achieve smooth tones. The coating permits scratching out of fine details such as hairs and small highlights. Such a pencil drawing must be reproduced by the halftone process.

One classic type of coated stipple board is Ross board, which was widely used until the 1960s when its manufacturer died. Since then, similar boards have come and gone. Such a board is now available from Medical Models Laboratory in Baltimore, Maryland and from British process boards (Ess Dee) in London, England and Dallas, Texas. These coated boards crack easily and must be mounted and protected like scratchboard.

Gesso

Paintings done with acrylic or oil are sometimes applied to a gesso-coated surface. Traditional gesso is made from calcium carbonate plus binders. A newer gesso is composed of acrylic polymer latex with white pigment, although a gesso-coated surface need not always be white. It can be tinted to blend with the painting. The grounds used may be canvas, illustration board, masonite, or wood.

As demonstrated by "old master" paintings done on gesso, this material ages well, but may crack if bent and should be protected like scratchboard.

Illustration Board

Illustration board is a relatively thick board that seldom requires mounting. It consists of a laminated series of three layers: the face layer, which is drawing paper; a central core of gray or white board; and a backing sheet which prevents curling or warping. In some brands this backing sheet is also drawing paper, thus providing two surfaces for drawing. Heavy weight or "double thick" boards have a heavier central core than the standard weight or "single thick" boards. Standard weight is more or less equivalent to a 14-ply paper, whereas heavy weight approximates 28 ply. However, illustration board is identified according to the press of the face layer which is either cold press (rough) or hot press (smooth). These terms refer to the temperature of the paper during its manufacture, when water is pressed out. Hot press papers proceed through rollers while hot and with a high water content, resulting in a smooth finish. Cold press papers cool and dry somewhat before being pressed and less water is removed, which provides a rougher texture.

Bristol Board

Bristol board is actually a paper and comes in weights of 1 to 5 ply. All the layers or plies consist of high grade bleached pulp or cotton; there is no central core of a different material. Bristol board bends easily if not mounted. Finishes are rough (vellum, medium or kid) or smooth (high or plate). Both finishes are generally smoother than their counterparts in illustration boards but are manufactured similarly. Plate finish looks somewhat shiny when curved under a light and is preferred by many artists for pen-and-ink line work.

Plastic Films

Some plastic and acetate films made of polyester and cellulose triacetate are frosted for drawing and used instead of paper, since they have the special advantage of being translucent. They can be scratched to achieve certain textures and the translucency eliminates the step of transferring a sketch, as is necessary when the final rendering is done on an opaque surface. The drawing can be traced and rendered in one step, which saves time.

Drawing films are usually coated with acrylic or abraded to produce a matte finish. Dupont's Cronaflex was originally manufactured for photographic use and is popular with many illustrators, since it has a thicker coating and better tooth than other drafting films. Acetate is a wood product and tears rather easily, but taping its edges helps prevent rips. Polyester films are stretched during the manufacturing process to increase strength and flexibility and will not tear, but they will twist or bend. Specially formulated inks and erasers are necessary for work on smooth or uncoated films, although media suitable for paper often will also work on coated films. Hard erasers damage coated film surfaces, yet water effectively erases ink from many films.

Tracing Paper

Transparent or translucent tracing papers, like plastics, permit tracing the original sketch in ink. Tracing papers vary in thickness, translucency, and durability over both the short and long term. They tear easily and some are slightly greasy as a result of their manufacturing process. "Vellum" is sometimes used as the generic term for all tracing papers, but true vellum is actually calf skin. Two specific types of drawing papers that most resemble true vellum are natural "vellum" papers and "vellum" parchment. Natural "vellum" is more transparent and more costly. "Vellum" parchment may also be opaque and may have a mottled, antique surface.

Coquille Board

Coquille board, like Bristol board, is not a board but a paper. It is approximately 3 ply in weight and is constructed like bristol board, but is embossed by rollers which produce a pebbled surface of various degrees of roughness, from fine to coarse. The finest texture is usually best for scientific drawings, since it permits showing finer detail (See Fig. III–15). Coquille is softer than Bristol and tends to catch pen points and tear slightly, which makes a clean line difficult to achieve.

Ledger Paper

Carbon dust drawings are sometimes rendered on ledger paper. This surface is constructed like bond paper but is heavier and stiffer, available in weights from 24 lb. to 44 lb. It is formed by the cold press method but has a smooth, matte "kid" finish. Like bond, ledger paper can be made from all-wood pulp, with various rag percentages, or from 100% cotton. Therefore, its inherent acidity and durability vary. Ledger paper should be mounted on a board.

Bond Paper

A few drawings may arrive at the editor's desk on bond paper, especially those from scientists who attempt to do their own illustrations. Bond paper is machine made like most other papers and may bear a simulated watermark impressed by a papermaking roller. As long as the watermark is white on white it will not show in line reproduction. Smooth, mounted bond paper that accepts a clean ink line, labeling, or tape is satisfactory for illustration. Such papers with a high cotton or rag content may outlast more exotic ground materials.

Mounting

Illustrations prepared on treated films, vellum, bond, Bristol board, scratch board, or other fragile papers should be mounted on stiff stock, usually white cardboard or illustration board, to prevent folding, wrinkling, or tearing. Various materials can be used to attach non-archival illustrations to the mounting stock, such as rubber cement, Fasson, dry mounting tissue, Spray-mount, and special tapes. Household paste, glue and conventional tape can wrinkle or damage light weight papers. When thin or transparent materials are mounted the adhesive may show through. Therefore, before mounting original illustrations the compatibility of the materials to be used should be tested on sample pieces. If longevity of the artwork is a concern, nondamaging, archival hinge mounting materials should be used and all non-archival adhesives avoided.

Transparent overlays showing instructions to editors and printers should be attached along an edge of the mounting stock, preferably in the back, with tape.

There should be at least a one-inch border of mounting stock left around all sides of the illustration to facilitate handling without damage. This is sometimes called the "printer's thumb." To further protect the finished product, a sheet of sturdy, opaque paper (cover stock) should be taped to the back of the mount and folded over the front so that it lies flat across the illustration. This will prevent scratches, soil, or scuffs on the face of the drawing during mailing and handling.

Identification

No matter how well an illustration is prepared, it will be for naught if the product is poorly identified and subsequently lost. Authors and editors must make certain that all submissions are identified on each part of the illustration. Each layer of the work must identify the author's name, title of the article or book and chapter, and the figure number. Instructions to the printer must be written on the layers that require the printer's special attention. The artist's signature should appear on the base drawing and the artist's label on the back of the mount. The top of the drawing or photograph should be labeled on its front as "top" to prevent confusion.

If illustrations are not mounted, instructions or identification should never be written directly on the back, as pressure marks will be visible from the front. Separately prepared labels on adhesive tabs should be applied to the top.

Some printers stamp numbers or attach other identifying material above the original art close enough so that it will show in the negative. If the original art is not to be marked, a flap of white paper can be extended down within one-half inch of the top of the illustration image, with the other end taped on the back of the mounting board (see Fig. III–36).

Packing for Shipment

The test of a strong packing case for an exhibit is to push it fully loaded from a truck bed without damage. A good test of well packed artwork is to have it survive the impact of the falling exhibit case. It is always safer to overpack any artwork that is being delivered by common carrier. Artwork should be covered with sturdy paper taped to the back and folded over the front. The artwork is then placed between two pieces of corrugated cardboard that are slightly larger than the art. The direction of corrugations should be at right angles to each other to minimize bending. These are then wrapped with heavy wrapping paper and strong mailing tape, labeled, insured, and sent off. For extra protection against bending, the number of corrugated pieces can be doubled, being sure to keep them at right angles to each other.

Delivery of Artwork

Hand-carried delivery enables the recipient to examine the art and take formal possession in the artist's or author's presence. If time is not a factor, United Parcel Service is presently much more reliable and less expensive than the Postal Service. However, insurance is not now available from UPS for artwork. Use of the Postal Service permits the package to be fully insured and a choice can be made between regular and express mail. Insurance on a valuable illustration may be expensive but may also provide necessary compensation for hours of work and illustration fees if the package is lost. Couriers can be costly, but when tight deadlines and reliability are important factors a good courier service is well justified. The package receives less handling, and an established courier is accountable.

The address of both sender and recipient must be complete and legible. Clear tape laid over the address label will prevent smears and running ink if the package gets scuffed or wet.

CHAPTER X
QUALITY CRITERIA, STANDARDS, AND INSTRUCTIONS TO AUTHORS

Authors usually communicate directly—face to face—with the illustrators who work with them, and journal editorial and production staff members know and communicate directly with each other and with their printers. The least direct link in the chain of communication described in Chapter 1 is the link between authors and the editors of the publications to which they submit their work; the first contact between editor and author is usually through the publication's "Instructions to Authors."

Because journal publishing offices follow a wide range of practices and procedures, and because printers may have different preferred ways of handling illustrations, it is most important that instructions for preparing illustrations be specific and detailed, so that authors and illustrators will know what is expected of the illustrations they submit. Keeping this in mind, editors should ensure that the "Instructions to Authors" provide answers to a number of important questions. The principal questions to be answered are (1) What kinds and how many illustrations are acceptable for publication? (2) In what form should they be submitted? (3) What should accompany them? (4) What else is the author responsible for? (5) Who can be contacted for answers not given in the Instructions to Authors?

Of course, not all of these requirements are appropriate for all journals, and those that are not should be modified or omitted. The requirements that follow reflect what many experts consider the best practices, in a variety of disciplines, for achieving the best results with scientific illustrations when printed. The following sections are cross-referenced to parts of the book that contain full details concerning the procedures in question.

Model Instructions to Authors

Examine issues of the journal to see what kinds of illustrations are appropriate and to estimate how many illustrations are considered reasonable in relation to the amount of text. Use illustrations only when they are needed—for efficiency and emphasis—to present evidence in support of the conclusions of the paper. Charts, graphs, and other line art should be used sparingly and only when a specific point must be illustrated; continuous-tone illustrations (for example, photomicrographs, photographs, and some drawings) should be used only when they are essential to the clarity of the presentation, since they require different paper or other special treatment. Arrangements must be made with the editor for excessive numbers of illustrations or for color. Because of the expense of color printing, illustrations will be published in color only when color is an integral part of the data and when the need for it can be substantiated. All or part of the expense of making color separations and printing color illustrations may have to be borne by the author (estimates of the cost of color reproduction should be obtained from the editor or publisher).

If possible, orient illustrations vertically rather than horizontally. Except for complicated illustrations showing large amounts of data, illustrations will probably be reduced to 1-column width (for example, on an average 2-column page of 8¼ × 11 inch trim size, 1-column width would measure 21 picas, 3½ inches, or 8.9 cm) or less, but column widths vary and so must be measured.

Titles and detailed explanations belong in the legends for illustrations, not on the illustrations themselves; however, the various parts of a composite illustration should be clearly labeled—such as A and B—on the figure itself. Legends should be typed (double-spaced) in sequence on a separate page. Each legend should have an Arabic numeral corresponding to the illustration, a short title (caption), and enough detail to make the illustration understandable without reference to the text (unless a similar explanation has been given in another figure legend). When symbols, patterns, arrows, numbers, or letters used to identify parts of the illustration are not defined on the face of the illustration, they must be identified and explained in the legend. If patterns and symbols are repeated in a series of figures, define them only in the legend of the first figure of the series. For photomicrographs, the internal scale of magnification and the method of staining should be identified. The legend should not be considered an opportunity to write another "Materials and Methods" section.

See Chapter XI

If a figure has been published, cite the original source and submit written permission from the copyright holder to reproduce the material. If a figure is adapted from an already published figure or figures, that, too, should be acknowledged. Permission is required, *regardless of authorship or publisher*, except for documents in the public domain, and the source of even those should be cited.

If a photograph of a person is used, either the subject must not be identifiable or the picture must be accompanied by the subject's written permission to publish the photograph. Blinds over the eyes are not a substitute for permission.

Preparing and Submitting Illustrations

Line Copy

Graphs, charts, maps, and other statistical line art should be accurately plotted, professionally drawn, and lettered with standard typefaces and kinds of lettering (transfer type, scriber, computer generated, or typeset) on white paper or other drawing material. Typewritten or freehand lettering is not acceptable; drawing on graph paper is not acceptable.

SEE CHAPTERS III, IV, AND IX

Artwork created on a graphic computer should not be recognizable as such; lettering, design, and inking should measure up to the standard artwork drawn by a professional illustrator or graphic artist.

SEE CHAPTERS IV AND V

If the publisher prefers to redraw charts and graphs to achieve a uniform style for publication, rough drawings or computer plots from the author may suffice. Also, as graphic computers become standard equipment in publication offices, editors may allow or even require authors to submit drawings and data for the points on their statistical illustrations, rather than finished artwork.

Line art should contain only black and white lines or dots, with no shades of gray. The illusion of shading should be created by combining lines or portions of lines (dots) in stippling, crosshatching, or patterned screens. Labels (words, letters, numbers, arrows, and symbols) and leaders should be arranged so that they fit the vertical or horizontal format. They should not cover, cross, or obscure any of the lines of the central focus of the drawing. If a leader crosses a line or a part of a drawing, that part should be whited out to reflect an imaginary light source from the top left corner of the drawing.

SEE CHAPTERS III AND IV

Letters, numbers, and symbols on charts and graphs should be clear and even throughout and large enough so that when they are reduced for publication each will still be legible. The abscissas, ordinates, data lines, and symbols must be large enough so that they will be clear when reduced for publication and the letters and numbers after reduction will be the size of 8-point type (2.0 mm in height). The smallest part of the illustration must be discernible. Original artwork should be at least 50% to 100% larger than the final (printed) dimensions. Use a reducing lens or copy machine that reduces to check the final effect.

SEE CHAPTER III

The thickness of ruled lines is also vital for clear presentation of data. Data lines should be heavier than axis lines, which should be heavier than standard deviations. Grid lines should not be included unless they are essential.

SEE CHAPTER IV

Define patterns and symbols in a key on the figure if possible. If they must be explained in the legend rather than on the face of the figure, only the following standard symbols, easily available to printers, should be used to denote points of observation on line drawings: ●, ○, ▲, △, ■, □.

SEE CHAPTER IV

Related figures in a series should be consistent in format, lettering, symbols, screens and patterns, scale, size, and reduction. The same symbol should be used to represent the same data in all figures of a given paper. Symbols, abbreviations, terminology, and spelling on illustrations should be consistent with the text, and standard abbreviations for units of measure should be used throughout.

SEE CHAPTERS III, IV, VII, AND IX

When submitting a manuscript that includes line art illustration, authors can send sharp black and white photographic prints, usually 5 × 7 inches (12.7 × 17.3 cm) but no larger than 8½ × 11 inches (21.5 × 28 cm) instead of the original artwork. Xerographic copies of the artwork (marked "not for reproduction") may be submitted for use by manuscript referees. The "best" set of photographs (to be used for reproduction) should be so identified. Halftones should be on glossy surface (F surface) paper; line cuts should be on semi-gloss paper.

The author should ask the publisher to use the original artwork of complex line drawings and tone illustrations for production when the manuscript has been accepted and is ready to go to the printer.

Each copy of each figure should have a label indicating the figure number, the last names of the authors, and the top of the figure. The information should be visible from the front so that it will appear on the negative and all subsequent proofs. Do not write on the back of artwork or photographs WITH ANYTHING. Photographs that are to be grouped may be mounted on flexible ground but must not be mounted on heavy cardboard. If parts of the composite are submitted separately, they should be accompanied by a sketch showing the desired placement. Do not combine line and tone art in one illustration. Do not use paper clips or staples on photographs or artwork. Do not bend or fold figures. Enclose each set of figures, protected by cardboard, in a separate envelope.

Continuous-tone Drawings

SEE CHAPTERS III, VII, AND IX

The artist may use a variety of media and grounds in preparing continuous-tone drawings, and these illustrations may be mounted (carefully) on stiff stock, usually white illustration board, to prevent folding, wrinkling, or tearing. Transparent or translucent flaps (overlays) with instructions to editors and printers should be attached along the top edge of the mounting stock with tape, or folded over the top and taped to the back.

Continuous-tone Photographs

SEE CHAPTER VII

The closer the reproduced image for printing is to the original image, the better the chances are for a good reproduction. Each step (generation)—internegative, photographic print, halftone negative, printing plate, ink image on press blanket, and ink dots on paper—that the image passes through decreases detail.

The preferred way to prepare a black and white tone photograph for publication is to start with an original negative on black and white film. A black and white contact or projection print is then prepared for submission to the editor or publisher. If a black and white reproduction must be made from an original image on a color slide, intermediate negatives should be made on panchromatic film (preferably 4 × 5 inches), with exposure and processing techniques customized to the slide. If it is a full-range slide, contrast reduction should be used. Common faults in prints prepared directly from color slides are high contrast devoid of details in highlight, shadow, or both, and highlights that have grayed down too much.

See Chapter VII

Submit the required number of photographic copies of continuous tone artwork as sets of glossy prints. Mark the best set "use for reproduction," or "best set." Do not submit xerographic copies.

Labeling for Continuous-tone Artwork and Photographs

Labels for continuous-tone artwork or photographs, such as arrows and letters, should be put on a transparent overlay, not directly on the work itself. The style and size of labels and arrows should be the same throughout the manuscript. Ideally, sizes should be consistent in the finished, printed work, so larger labels should be used on artwork that will be reduced more than others. Labels and arrows must be explained or identified in the figure legends.

See Chapters III and VII

Grouping

Pictures can be grouped as parts A, B, and C, for example, of one figure; grouped parts of figures that are the same size are easiest and least expensive to reproduce. Pictures to be grouped should usually be submitted as individual prints, not cut and mounted, to allow leeway in planning the best layout. Some publishers, however, may specify that photographs be mounted on thick, flexible, smooth-surfaced drawing paper. A multi-part figure of matched photographs (a composite) should be mounted so that an even amount of space separates each print within the plate or so that they abut. If several photographs are to be combined in a multi-part illustration, or if a special layout is desired, a detailed sketch and instructions should be submitted. Avoid combining line art and tone art in the same group.

See Chapter VII

SEE CHAPTERS III AND VII

Tissue-paper or acetate overlays are used with photographs to indicate cropping, position of labels, and important parts of images (so that details can be matched exactly or enhanced). Marking should be done with a very soft pencil (or grease pencil on acetate) with extreme care, so as not to damage the artwork or surface of the photograph beneath.

Photomicrographs

Prints of photomicrographs should be submitted in the exact size for publication; trim prints to leave only the area to be published and add internal scale markers. Symbols, arrows, or letters used in the photomicrographs should contrast with the background (white on black, or black on white) and should be explained in the legend. All these marks should be put on a transparent overlay, not directly on the print. Some publishers, however, will instruct the author to affix these marks to the surface of the print. This requires great care to avoid disfiguring the print emulsion.

Electron micrographs should be covered with a transparent overlay; the critical lines or areas that must be reproduced with greatest fidelity should be indicated on the overlay. Care should be taken to avoid marring the photograph when marking the overlay.

Tracings

Original electrocardiographic and other tracings should not be submitted because they are easily damaged by heat and pressure. Provide instead a photomechanical transfer (PMT) print which will also enhance the image; pale gray lines will appear black and the background white.

Color Illustrations

Authors should consult with the editor or publisher about the use of color when submitting their manuscripts. Although color may in some instances be essential to the documentation of scientific results or proof, it should seldom if ever be used for esthetic or decorative purposes, since the printing process is expensive and the cost must usually be borne by the author.

Spot color

SEE CHAPTERS III AND VIII

Spot color can be used to highlight a particular segment or segments of a black and white illustration, particularly in line illustrations such as diagrams, charts, graphs, and maps. A separate overlay showing where color is to be added is prepared for each color. Spot color also can be used with a black and white halftone to produce a duotone.

Process color

See Chapter VIII

The four-color reproduction process brings the full spectrum of color to the printed page. Artwork for process color treatment is prepared in one of two ways: as reflection copy (color photographic prints, drawings and paintings, and graphics) or transmission copy (color transparencies, such as 35-mm slides). Positive color prints are preferred, because authors, editors, and printers will find it easier to match proofs against prints than against transparencies. Prints are more easily cropped and assembled into composites (plates), and overlays can be applied to carry symbols, lettering, and scale markers.

Color transparencies (if requested by the publisher) should be protected by a transparent sleeve of appropriate size. The cardboard mount should be properly labeled with the figure number, the author's name, and an indication of the top margin. Crop marks and labeling should be indicated on a separate diagram or on an enlarged color print. If transparencies are to be grouped into a composite or plate, an accompanying diagram should indicate the position of each component. Duplicate transparencies or color prints should be submitted for use by the reviewers.

Color prints should carry a label on the back with the figure number, the author's last name, and an indication of the top margin. (Note: some printers prefer to have the label on the front.) Figure numbers, symbols, and scale markers can be placed on the surface of the print or on a transparent overlay (the latter is preferred). Prints for a composite color illustration should be cropped and mounted so that an even amount of space separates each print within the plate. This space should measure 1/16 to 3/16 inch (2 to 5 mm) between adjacent prints. Use thick, heavy-weight, highly flexible, smooth-surfaced drawing paper for mounting. Do not mount prints on stiff cardboard. Submit duplicate sets of color prints to assist editors and reviewers in their evaluation.

CHAPTER XI
LEGAL AND ETHICAL CONSIDERATIONS

The legal aspects of scientific illustration derive from ethical concerns of authors, illustrators, publishers, and printers. Legal considerations include matters of copyright, copyright release, permits for clinical photography of human subjects, contractual arrangements for works made for hire, and the rights and liabilities of illustrators, authors, editors, and publishers.

The ethical aspects are rarely codified into a set of rules for standards of ethically proper conduct (normative ethics), but certain guidelines do exist for ensuring fair practices among the various professions responsible for the publication of scientific illustration.

Copyright

The laws of copyright are designed to protect the rights of those who create communications. The rights imply ownership of the work, control of its use, and entitlement to any financial benefit that might derive from its use. Because of the great difference in commercial potential between scientific journal and scientific book publishing, copyright protection is financially more important for the illustrators, authors, and publishers of books. Although the copyright requirements for journal and book publication are identical, the financial implications of transfer of ownership, reuse or republication rights, foreign translation and publication, and other copyright matters are more significant for book publishing.

In the United States, the Federal Copyright Revision Act of 1976 (Public Law 94-533) broadened the definition of copyrightable materials and extended the limits of protection for the owner (or holder) of the copyright. All of the graphic forms described in this book are copyrightable, but the owner of the copyright could be any of the persons involved in the chain of creation and production that leads to publication.

Copyright confers on its holder the ownership of exclusive rights to copy, distribute, sell, publish, and reprint the work, in similar or modified form. These rights apply to all forms of publication, including collective works (periodicals or journals). The absence of written agreement between the creator and the publisher of the work does not prohibit publication in a periodical, since section 201 of PL 94-553 states, "In the absence of an express transfer of the copyright or of any rights under it, the owner of the copyright in the collective work is presumed to have acquired only the privilege of reproducing and distributing the contribution as part of that particular collective work, any revision of that collective work, and any later collective work in the same series" (Public Law 94-533).

Copyright information in this chapter is based on the copyright law of the United States, but international protection is guaranteed for copyrighted works from all countries subscribing to the Universal Copyright Convention. Most countries are members of this Convention and/or the Berne Convention, both of which recognize copyrights originating in other member countries. Although the United States is not a member of the Berne Convention, its membership in the Universal Copyright Convention provides protection for almost all works published in the United States. This agreement stipulates that a member nation must provide the same copyright protection to other member nations that it guarantees to its own authors. Copyright requirements are described further in the current edition of the "CBE Style Manual" (Style Manual Committee, 1983) and in a number of recent publications (Crawford, 1977; Crawford, 1978; Latman, 1979; Quintiliano, 1978; Strong, 1981; Wincor, 1982).

Terms of Copyright

For works published after January 1, 1978, the term of copyright is the lifetime of the author or creator plus fifty years. An exception is made for works that are created as part of an employee's job responsibility or that are commissioned for certain forms of publication by written agreement with the author or illustrator. Such works are designated "work for hire"; their stipulated term of copyright is seventy-five years from the date of publication or one hundred years from the date of creation, whichever is shorter.

Transfer of Copyright

In scientific publication, publishers usually request that illustrators and authors transfer their respective copyrights. This process can also be initiated by the author or illustrator. Transfer is through a written contract or agreement, which states the conditions of copyright transfer. Authors and illustrators are afforded additional protection under the law, since they or their heirs are permitted to terminate the transfer agreement within specified periods of time after publication.

The collective nature of much scientific publication may obscure the issue of authors' and illustrators' rights. In journal articles, for example, the process of scientific illustration often begins with the data produced by the author/scientist. The illustrator serves as a reviewer and consultant on how the data may be reproduced for publication. After preparing the illustrations (usually for a fee in the case of a freelance illustrator), the illustrator transmits the work to the author, who incorporates it into the scientific article. The author then submits the article to a journal editor for peer review. After review, the article is usually returned to the author with specific recommendations for revision of the text and illustrations. If the revised version is accepted for publication, the article is prepared by the editorial office or the publisher for printing. Typesetters, production photographers, and printers complete the process, which ends with the distribution of a periodical containing a collection of similarly produced articles. In book publication, much the same process may be followed.

At some time before printing, the editor or publisher will ask the author to transfer copyright. Ideally, this transfer should take the form of a publishing agreement in which the author transfers ownership of the work to the publisher and the publisher agrees to publish it in a specified manner. The agreement or contract may stipulate the author's obligations (originality of the work, participation of co-authors, permissions to include previously published material, and so on) and the publisher's responsibilities (for example, name and format of the publication, handling of proofs, right of the author to reprint work in another format).

The rights of the freelance illustrator are often overlooked in such agreements. Publishers commonly assume that the author of an illustrated journal article has copyright for all components of the work and can therefore assign all rights to the publisher. Authors and editors may assume that the illustrated material is covered by the copyright, since the illustrator received a fee for creating it. The freelance illustrator, however, may still own the copyright to the illustrations unless a written agreement transferring ownership to the author (or editor) was executed. Such an agreement would designate the illustrations as "works made for hire" (see below). Illustrations produced under such an agreement between illustrator and author would be legally owned by the author. Similarly, work created by illustrators as part of their employment responsibility would be owned by their employer. This is the usual situation when the illustrator is employed in a biomedical or graphic arts department of an academic or commercial institution. Charts, graphs, drawings, and photographs may be prepared for institutional authors for use in their scientific reports. Copyright of such works technically belongs to the institution (the employer), but the author (as an agent of the institution) is usually privileged to assign copyright to the publisher. This policy may vary, however, and authors and illustrators may have to comply with the copyright policy of their own institution.

Before the scientific journal article or book is committed to publication, the editor must obtain an agreement for transfer of copyright or for specific rights to use the work in published form. Before signing the transfer agreement, the author should be certain that the conditions of ownership have been satisfied:

- Is the work (including its illustrations) original?
- Have permissions been obtained for use of previously published material, and has appropriate notice of permission been included in the manuscript?
- Are the illustrations owned by the illustrator?
- If illustrations were commissioned, was written agreement obtained defining the illustrations as works made for hire, or was copyright transferred?

Section 105 of the 1976 Copyright Act stipulates an important exception to this process. Authors of articles prepared as part of their "official duties" as employees of the United States Government are unable to transfer copyright. Their articles are in the "public domain." Copyright transfer agreements should recognize this.

To cover such circumstances, many scientific journals provide a copyright transfer agreement that encompasses several contingencies. The agreement may include provision for assignment of copyright from an independent author, a non-government employer of the author, or an author who is employed by the United States Government:

Copyright Assignment

of an article to be published in the (..name of publication..), entitled (..title of manuscript..):

1. I affirm that the written and illustrated material in the above manuscript has not been previously published and that I (and any co-authors) own and have not transferred elsewhere any rights to the article.

or

2. I affirm that I have obtained written permission to use any previously copyrighted material included in the article and that such documentation will be forwarded to (..publisher..).

and

3. I hereby assign and transfer to (..publisher..) all exclusive rights of copyright ownership of the article (including the rights of reproduction, derivation, distribution, sale, and display), as pro-

tected by the laws of the United States and foreign countries. These exclusive rights will become the property of (..publisher..) from the date of acceptance of the article for publication in the (..name of publication..). I understand that (..publisher..), as copyright owner, has sole authority to grant permission to reprint the article.

4. I sign for and accept the responsibility for transferring copyright of this article to (..publisher..) on behalf of any and all authors.
_________________________ (signature of author)
_________________________ (date)

or

5. I prepared the article as part of my official duties as an employee of ____________________ (name of company or organization).
____________________ (name of author).
I, as a duly authorized representative of the company or commissioning organization, hereby transfer copyright of the article to (..publisher..).
____________________ (name of authorized agent)
____________________ (date)

or

6. I prepared the article as part of my official duties as an employee of the United States Government. I am therefore unable to transfer rights to (..publisher..).
____________________ (name of author)
____________________ (date)

Work for Hire

Each type of artwork discussed in this book is the property of the illustrator and is covered by his or her copyright unless the work falls under the definition of "work made for hire." The definition covers work produced under either of two conditions: (1) work produced as a specific condition or product of employment and (2) commissioned or specially ordered work of certain types. The latter includes contributions to collective works (such as scientific journals), translations, compilations, and various supplementary works. Not all commissioned works qualify for this definition, and those that do require a written agreement between the illustrator and the commissioner, specifying them as works for hire.

The contractual rights of the illustrator, author, editor, and publisher can be assured only if the conditions for the use of illustrated material are considered. If the artwork was not made for hire as defined above, the assignment or transfer of copyright from the author may be exclusive but limited. Such a statement of rights may stipulate limitations on the use of the work (reprinting, translation, and so on), the duration of the transfer, the conditions for termination of the transfer, the form of copyright notice, and other conditions that affect publication. Limited transfer of rights to publish illustrated material is more likely to be associated with publication of books, since journal articles are usually limited to a one-time publication as part of a collective work. Provisions may have to be made by the publisher to return the artwork to the author or illustrator after publication. Instructions for the return of manuscript and illustrations should be included in the transfer agreement. In most instances, the responsibility for developing the necessary agreements for the publication and subsequent use of illustrated materials rests with the author and illustrator.

The Doctrine of Fair Use

Copying and quoting of brief portions of a copyrighted work without written approval from the copyright holder are permissible under certain conditions. Although the definition of these conditions is not precise, limited copying of copyrighted material for scholarly and nonprofit educational purposes may meet the test of fair use. For example, single copying of a published work for preparation of a lecture is permissible, but multiple copying for student use in a classroom is not if such use would detract from the economic value of the work for the copyright holder. The four criteria for determining whether or not a copyrighted work may be copied without permission are (1) the purpose and character of the use (for example, for commercial or nonprofit educational use), (2) the nature of the work copied, (3) the amount and substantiality of the portion used in relation to the copyrighted work as a whole, and (4) the effect of such use on the potential market for or value of the copyrighted work. These criteria are relevant to both scientific journal and book publishing.

Permission to Reprint

Scientific authors are frequently invited to contribute to collective works of a secondary type (such as reviews, symposia, or conference proceedings). These publications borrow heavily from previously published works by reprinting charts, graphs, drawings, and photographs. Authors may be unaware of the necessity to request permission to reprint illustrations from their own previously published work, whether or not they have formally assigned the copyright to an editor or publisher. Written permission to reprint must be obtained from the copyright holder and appropriate acknowledgment of prior publication must be displayed in the secondary publication. This is the author's responsibility.

The author must first determine who holds the copyright on illustrations to be used. This might be another author, an illustrator, a scientific society, or the publisher. Information about copyright is customarily displayed in the front matter of each issue of a journal. A letter requesting permission to reprint must then be directed to the copyright holder. The letter should identify the primary publication (for example, journal title or book), the author, article title, date of publication, and the pages on which the material appears. Figure numbers should be specified. The secondary publication in which the reprinted portion is to appear should be identified by stating the author or editor and the title. Some indication of the publication date should be included, and some copyright holders may request information on the list price of the secondary publication and the number of copies to be printed (the press run).

If the illustration is to be modified in the secondary publication, details of the modification must be included. Permission to alter the illustrations may require the approval of the illustrator, depending on the type of copyright ownership involved. Whatever the arrangement, the author should consult with the illustrator about any design changes.

A credit line or acknowledgment must be published for each portion of a reprinted illustration. The credit should identify the author(s), title of the source article, title of the publication (journal or book), volume number (for serial publications), initial or inclusive page numbers of the article, name of the publisher (for books), and the year of publication. The content or specific wording of the credit line may be supplied by the copyright holder. An example of a credit line for a reprinted illustration might be:

> Figure #. (Reproduced, with permission, from Rubinstein NA, Kelly AM: Development of muscle fiber specialization in the rat hindlimb. J Cell Biol 90:128, 1981.)

Permits in Clinical Photography

As in most clinical procedures involving human patients, clinical photography requires the patient's permission. The authorization to photograph must be an informed consent that explains the procedure, the purpose and use of the photographs, the method for preserving anonymity, and the right of patients to withdraw permission at any time before publication without prejudice to their treatment. Is informed consent required for photographs taken during the course of an operation or of portions of the body in which the identity of the patient is not revealed? The procedure of informed consent should be followed, because clinical photographs are often displayed publicly in the form of projected slides, exhibits, or published illustrations. Permission is also needed to reproduce the likeness of an individual in a drawing.

The photographer must obtain the necessary permits before the subject is photographed, but the author must be able to verify that photographs for publication were taken with the informed consent of the subject. Explicit verification is not usually required by the publisher in the copyright transfer form signed by the author. In signing the form, however, the author asserts that the manuscript contains no material that defames or invades the privacy of other individuals; the author also indemnifies the publisher and assumes responsibility for any legal claims that might arise from publication of clinical photographs. Sensible guidelines for the development of informed consent in clinical photography are provided by Hansell (Hansell, 1979), Tarcinale (Tarcinale, 1980), and Reynolds (Reynolds Jr., 1985).

Responsibilities and Liabilities

Each person in the chain of publication assumes specific responsibilities that affect the transfer of graphic materials from the illustrator to the printer. Each, in turn, incurs certain liabilities if the process fails.

Illustrators

A verbal or written contract usually exists between the illustrator and the buyer of the illustrations. The contract assumes that the illustrator's product is original, that none of its content is plagiarized, and that no copyrighted material is included without permission. The illustrator is also obligated to produce the contracted work within a stipulated time. Failure to meet these conditions would be grounds for voiding the contract and for canceling of fees. If illustrations containing plagiarized material or that copyrighted by someone else are published, the liability for such violations could extend to the author and publisher.

The illustrator should reach explicit agreement with the buyer (for example, author, editor, or publisher) concerning the use and ownership of the artwork and its return after publication. The freelance illustrator should also provide a written guarantee to the buyer that the work will not be sold competitively while publication is in progress or within a specified period of time that coincides with the active sales of the published work. These agreements are particularly applicable to book publishing.

Authors

It is usually the author who must meet the cost of producing the original illustrations. In spite of this, the author must recognize that ownership of the illustrations may remain with the illustrator unless a written agreement states otherwise. The author may be obliged to return the artwork after publication, and the editor or publisher should be informed of this condition at the time the manuscript is submitted for their consideration.

Provisions for protecting the artwork against damage or loss should be made while the work is in the author's custody. The work must also be protected during transport to the editor or publisher through the use of support materials, overlays, and appropriate packaging. Just as the author would not alter the text of a scientific communication without consulting the co-authors, artwork should not be altered without consulting the illustrator.

In submitting a manuscript for publication, the author asserts that the data are original. Because a scientific illustration is part of the data, the assertion of originality extends to this component of the manuscript. No form or degree of plagiarism can be accepted by the editor, and no previously published illustrations can be used without the necessary permissions and credits. If such material is published, liability is shared by the author, editor, and publisher.

Unintentional violations of copyright are not infrequent. Authors may incorporate illustrations from their own published articles assuming that, because they created or commissioned the original work, it is their right to reproduce parts of it in a new manuscript. The previous copyrights would, of course, be violated if the new manuscript were published without the required approvals. The author's use of the copyrighted illustrations might also raise some question about the originality of the "new" manuscript, unless the author provides explicit reasons for doing this.

Editors

As the next custodian of the manuscript, the editor must guarantee the security of the artwork and must assume liability for damage or loss during the review process. Particular care must be taken to avoid marring or disfiguring the surfaces of drawings and photographs by careless handling of the manuscript. No alterations should be undertaken without permission of the author and illustrator. Permission must also be obtained to use the artwork for any publication other than that for which the work was originally intended.

The editor must be especially attentive to the methods of transporting manuscripts; transport includes forwarding to referees for review, return to the author for revision, and transmittal to the publisher. The editor must respond to the author's request for return of artwork after publication by making the necessary arrangements with the publisher.

Publishers and Printers

After the manuscript is committed to the publisher and printer, the same safeguards against damage and loss must be observed. By contract, the publisher has agreed to publish the illustrations in a manner acceptable to the illustrator, author, and editor. The publisher may not do this without receiving a transfer of copyright from the author or an appropriately signed exemption. The transfer may involve all or only part of the rights. Normally, the printer returns the artwork to the publisher after the work is printed. Subsequent storage of the artwork for a certain length of time must be under optimal conditions. The publisher must then return the illustrations to the author or illustrator, according to the conditions of the publishing agreement.

The publisher may not reprint the illustrations or use them in another published form without written permission from the copyright owner unless all reproduction rights were transferred originally. Conditions for reprinting may be explicitly set forth in the copyright transfer agreement.

The printer must guarantee protection of the artwork during all phases of production. This may include sizing, photographing (halftones, color separations, and so on), stripping and imposition, platemaking, printing, cutting, binding, and distribution. After the artwork is returned (usually to the publisher), the printer's responsibility ends. The printer is not required to retain any of the materials used in printing (for example, negatives or plates), but storage for a limited time is often provided as a convenience for the publisher. The stored plates can then be used for reprinting with a certain amount of cost saving. The responsibilities of the printer in the publication process are described in "A Guide to Writing and Using Printing Contracts," published by the Printing Industries of America (Mattson, 1977). This publication contains models of printing contracts of various complexities and specificities.

Ethical Guidelines

The foregoing catalog of professional obligations, by no means comprehensive, provides a basis for normative ethical standards in the management of scientific illustration. Like most codes of professional conduct, obligations are perceived as absolute (not conditional) and performance standards are described in terms of what is inherently assumed to be the "right thing to do" (not of the consequences of the action itself). Published guidelines are few. Notably, the Graphic Artists Guild publishes a Code of Fair Practice, governing the relations between the illustrator and the customer (Graphic Artists Guild, 1984). This code emphasizes the obligations and rights of the illustrator in creation and use of illustrated material. The Guild has established a mechanism and a grievance committee to review abuses of fair practice.

The Institute of Incorporated Photographers (Institute of Incorporated Photographers, 1969) has published a Code of Professional Conduct that sets standards of behavior for their membership in the context of relationships between employer and employee and among members themselves. Turnbull has contributed a thoughtful chapter on the ethical practices of the medical photographer, covering the relation of the photographer to the medical profession, the patient, and the hospital (Hansell, 1979), and Tarcinale has emphasized the responsibility of the photographer to protect the patient's right to privacy (Tarcinale, 1980).

The ethics of scientific communication have been discussed extensively (Cohn, 1960; Price, 1964; Wright, 1970), but authors, editors, and publishers have yet to develop formal statements of ethical conduct. The growing concern of authors and editors about duplicate publication of scientific results, fragmentation of research reports into multiple publications, excessive multiple authorship, and fraudulent conduct of research has contributed to a reexamination of our standards for scientific research. Huth has defined specific rules for authorship (Huth, 1982), and he has discussed the ethical responsibilities of authors, editors, reviewers, and publishers in the current edition of the "CBE Style Manual" (Style Manual Committee, 1983).

The Association of American Medical Colleges has issued a report on ethical standards in research (Association of American Medical Colleges, 1982). The report recommends nine procedures for faculties and institutional officials to follow in their effort to reduce research fraud. Two of these are directly concerned with scientific publication. They call for "examining institutional policies on authorship of papers and abstracts to ensure that named authors have had a genuine role in the research and accept responsibility for the quality of the work being reported," and "examining the institutional role and policies in guiding the faculty concerning public announcement and publication of research findings." In the event that an alleged fraud is substantiated by institutional investigation, the report further recommends that "all pending abstracts and papers emanating from the fraudulent research be withdrawn and editors of journals in which previous abstracts and papers appeared should be notified." The act of publication presupposes that what is published is accurate, honest, and original. It follows that the graphic materials contained in such publication are presumed to be true representations of the author's knowledge of the subject. The reader relies on the editorial and peer review process to guarantee that these conditions of scientific publication have been met. Nevertheless, ultimate responsibility for truth lies with the author and the illustrator. The editor, publisher, and printer share this responsibility by providing the most accurate translation of the original illustration into printed "scientific truth." Successful and ethical performance of this task requires that each participant in the process honor the rights, and acknowledge the responsibilities, of the illustrator, author, editor, publisher, and printer.

LITERATURE CITED

Association of American Medical Colleges: AAMC ad hoc committee on the maintenance of high ethical standards in the conduct of research. J Med Educ 57:896–902, 1982

Crawford TC: Legal guide for the visual artist. New York, Hawthorn Books, 1977

Crawford TC: The visual artist's guide to the new copyright law. New York, Graphic Artists Guild, 1978

Cohn V: Science writer ethics. National Association of Science Writers Newsletter, 8:11, 1960

Fetzner JM: Copyright transfers: developing policies and procedures. CBE Views, 2 (no. 4):8–9, 1979

Graphic Artists Guild: Pricing and ethical guidelines. New York, Graphic Artists Guild, 5 ed., 1984

Hansell P: A guide to medical photography. Baltimore, University Park Press, 1979

Huth E: Authorship from the reader's side. Ann Int Med 97:613–614, 1982

Institute of Incorporated Photographers: Code of professional conduct. Ware, Institute of Incorporated Photographers, 1969

Latman A: The copyright law. Washington, D.C., Bureau of National Affairs, 5 ed., 1979

Mattson GA (ed.): A guide to writing and using printing contracts. Arlington, Va, Printing Industries of America, Inc., 1977

Price DJS: Ethics of scientific publication. Science 144:655, 1964

Public Law 94-553, enacted October 19, 1976. Title I—General Revision of Copyright Law. Superintendent of Documents, U.S. Government Printing Office, 1977

Quintiliano B: The copyright revision act and problems of journal publication. CBE Views 1 (no. 2):8, 1978

Reynolds Jr. LR: Photography and the patient's rights. J Biol Photogr, 53:117, 1985

Strong WS: The copyright book: a practical guide. Cambridge, MIT Press, 1981

Style Manual Committee: CBE style manual. Council of Biology Editors, Bethesda, MD, 5 ed., 1983

Tarcinale MA: Medical photographer's role in protecting a patient's right to privacy. J Biol Photogr, 48:183, 1980

Turnbull PM: Ethical considerations. In Hansell P: A guide to medical photography. Baltimore, University Park Press, 1979, Chapter 11, p. 147

Wincor R: Rights contract in the communication media. New York, Law and Business Inc., Harcourt-Brace-Jovanovich, 1982

Wright RD: Truth and its keepers. New Scientist 45:402, 1970

CHAPTER XII
GLOSSARY OF TERMS USED IN GRAPHIC ARTS

The glossary was assembled from various sources with the kind permission of the publishers. The following publications should be consulted for more comprehensive listings of terms that are commonly used in the graphic arts and printing industries:

Elsevier's Dictionary of the Printing and Allied Industries. F. J. M. Wijnekus, Elsevier Publishing Co., Amsterdam, 1967.
Graphic Arts Encyclopedia. G. A. Stevenson, McGraw-Hill, New York, NY, 2 ed., 1979.
Graphic Arts Glossary. Edwards Brothers, Ann Arbor, MI, 1978.
Phototypsetting. A Design Manual. James Craig, Watson Guptill Publications, New York, NY, 1978.
Pocket Pal. International Paper Co., New York, NY, 13 ed., 1983.
Production for the Graphic Designer. James Craig, Watson Guptill Publications, New York, NY, 1974.
The Lithographers Manual. Raymond Blair, Graphic Arts Technical Foundation, Pittsburgh, PA, 8 ed., 1988.

abscissa The horizontal line or X axis of a two dimensional coordinate grid, or any horizontal line parallel to the horizontal axis.
acetate A transparent sheet made of clear plastic; frequently used to make overlays.
acrylic Paint that looks like oil but is pigment suspended in an acrylic polymer emulsion mixed with synthetic plastic resins. As water evaporates the resin molecules combine, creating an adhesive, waterproof, flexible film.
adhesive binding See **perfect binding**
airbrush Small, hand-held sprayer used like a paintbrush to shade photographs or drawings. It can be used with inks or thin paints. It is powered by compressed air, hence its name.
algorithm A series of rules or steps to be followed in a specified sequence to solve a problem (often displayed as a chart or diagram).
Amberlith Amber-colored masking and stripping film coated on a polyester backing sheet. Amberlith is a registered trademark of the Ulano Company.
axis Fixed reference line for measurement of coordinates in graphs;

commonly, a horizontal line (X axis) intersecting at a specified origin with a vertical line (Y axis).

blanket A fabric coated with natural or synthetic rubber, clamped around the blanket cylinder, which transfers ink from the plate to the paper.

bleed Any part of the printed area (usually a photograph or non-illustrative artwork; never headlines or copy) that extends beyond the trim edge of the page.

blueline A paper print made from a single negative or a flat, used primarily as a proof, to check content and/or positioning; also called brownline, brownprint, blues, or Van Dykes.

bond paper Good quality paper often with a high rag content. It can be manufactured with many different characteristics including good erasibility, high ink penetration, etc.

Bristol board One or more plies of good quality paper made from bleached pulp or cotton rag. Available in kid or vellum (both slightly rough) or plate (very smooth and slightly shiny) finishes.

brownline See **blueline**

camera-ready copy Typeset or typewritten copy, together with line drawings and continuous-tone copy, that is ready in all respects for photomechanical reproduction. When copy is typeset, the printer refers to it as "etch proofs" or "reproduction proofs." Camera-ready copy may also be called "black and white" when continuous-tone copy is not included.

carbon dust technique An illustration technique in which carbon dust (also called "sauce"), prepared from sandpaper-rubbed carbon pencils, is brushed onto a drawing to obtain various degrees of shading.

case bound A term denoting a book bound with a stiff or hard cover; a hard-cover book.

casein A product of skimmed milk used for sizing and as an adhesive in the manufacture of coated papers; can be used in place of albumin for making surface plate coatings.

chart A form of graph in which the fluctuations of the magnitude of one variable, such as temperature, barometric pressure, price, population, etc. are represented.

coated paper Paper coated with white clay or an acrylic substance to provide a smooth printing surface. Coated paper is usually glossy but can also be dull.

cold press Describes a roughly textured paper surface resulting from pressing of paper after it has cooled and slightly dried.

collate To gather (assemble) sections (signatures) in proper sequence for binding.

color bars A series of colors imprinted on process color proofs to show the four colors used to print the image. The bars show the amount of ink and their densities across the press sheet.

color correction Techniques of improving color reproduction; includes matching, dot etching, filtering, and scanning.

color key papers Papers coated with colored inks from the Pantone Matching System (PMS) or other industrial ink colors. Use of these colors provides a fairly accurate idea of what the printed version will look like.

color proofs–pre-press Proofs made by photographic or photomechanical means; faster and less costly than press proofs.

color proofs–press Proofs prepared on a printing press in advance of the production run; more accurate than pre-press proofs, but more expensive.

color proofs–progressive Progressive proofs ("progs") consist of plates of each color used in printing the final image: yellow, cyan, magenta, and black. When superimposed, the plates will show the sequence of printing and the effect when each color has been applied.

color registration The accurate positioning of one color plate over another so that both are in correct relationship to produce the effect of a single image. If the plates are "out of register" the resulting image will appear fuzzy. Registration is facilitated by matching registration marks on each plate.

color separation The process of separating color into magenta, cyan, yellow, and black by photographic or scanning techniques. Progressive proofs are prepared from such separations.

color spot See **spot color**

column width The actual width of one column of type in a journal or book; usually specified in picas.

composition The assembling of characters into words, lines and paragraphs of text or body matter for reproduction by printing. Also the arrangement of elements in an illustration.

constant In statistics, a function with a fixed value (see **variable**).

continuous tone A photographic image that has not been screened and which contains gradient tones from black to white.

contouring Thick and thin lines in a drawing that depict tone and form (also called **eyelashing**).

contrast The density difference (total gradations) between the highlight and shadow areas of copy; also called **copy density range.**

coordinate surface A grid constructed of two axes, one vertical (Y axis) and one horizontal (X axis), which intersect at the origin to create four quadrants. The relation between two variables, X and Y, can be graphically plotted in a two-coordinate system. Three or more coordinates (X, Y, Z, etc.) can also be plotted.

copy Any material, including photographs, rules, designs, and text, that is used for reproduction by printing.

copyright Literary, dramatic, artistic and musical property protection for the creator as authorized by the U.S. Constitution, securing for the creator, for specified times, the exclusive rights to his or her work.

Coquille board A textured, uncoated paper available in several degrees of roughness.

Cromalin A laminated series of color-impregnated film used as a form of proof for color reproduction.

crop To opaque, mask, cut, or trim an illustration, or its reproduction, to the required size, to fit a specified area.

crop marks Marks placed at the margins of an illustration to indicate the area to be eliminated from the portion of the illustration to be reproduced.

cross-hatching Intersecting lines used to create tone and form.

curve Line, straight or curved, drawn on a coordinate surface to denote the relationship of two variables.

densitometer Photoelectric device that measures the density of photographic images.

density In general terms, the relative darkness of an image area as seen by the eye; in technical usage, a measure of light-stopping ability or blackening of a photographic image as read on a densitometer.

dot etching Changing tone values by chemically reducing the size of, or "etching," halftone dots. This method is used in lithography when tone values or color strength must be changed during the photographic steps rather than on the printing plate.

dot gain Slight enlargement of a halftone dot during photographic exposure or development, or during printing on the press.

drawing film Translucent or transparent materials made respectively from polyester or cellulose triacetate. Clear (transparent) films are usually used for overlays. Frosted (coated) films have microscopically rough surfaces which accept pencil and ink.

drop in An image or text inserted into another illustration.

drop out 1. (verb) Lost in reproduction, for example, loss of detail in gray areas of line copy during reduction. 2. (noun) The photographic procedure for eliminating part of an illustration, usually the background, while retaining other parts.

dual dot screen Halftone screen containing two dot patterns of different size, used to extend detail over a wider density range.

dummy pages 1. A preliminary drawing or layout showing the position of illustrations and text as they are to appear in the final reproduction. 2. A set of blank pages made up in advance of printing to show the size, shape, form, and general style of a piece of printing.

duotone A two-color halftone reproduction from a one-color original, requiring two halftone negatives for opposite ends of the gray scale, at specified screen angles. One plate is usually printed in dark ink, the other in a lighter one.

Dylux The Dupont system for making photographic proof from negatives.

elliptical dot Elliptically shaped dots on a halftone screen which give improved gradation of tones, particularly in the middle range.

emulsion A photographic term for a gelatin or collodion solution holding light-sensitive salts of silver in suspension; used as the light-sensitive coating on glass photographic plates, film, or metal plates.

engraving Incising designs or images on the surface of a material from which printing impressions can be made.

exposure A photographic process in which an image is produced on a light-sensitive emulsion by exposure to light.

exposures–bump A supplementary exposure used in shooting halftones, without the screen, to give greater separation of the highlight tones.

exposures–flash In halftone photography, the supplementary exposure given to strengthen the dots in the shadow areas of negatives.

exposures–main The camera exposure made through the halftone screen to reproduce in the negative all areas of a photograph except the deeper shadows.

fixative Any clear solution sprayed or coated on artwork or other material, such as reproduction copy, that "fixes," or stabilizes, the image, rendering it more resistant to wear or smudging.

flat 1. In lithography, the assembly of photographic negatives or positives on goldenrod paper, glass, or vinyl acetate, from which the printing plate is made. 2. In photography, a picture that is lacking in contrast.

folio Page number. Number of a page at top or bottom (for preliminary matter, Roman numerals in lower case).

font A complete assortment of type of one size and face, containing all the characters needed for ordinary composition.

form Any assembly of pages that can be printed simultaneously in a single impression of the printing press; a flat of imposed negatives.

format General form of a book, brochure, direct-mail piece, or other printed matter, with particular reference to composition, layout size, and general appearance.

four-color process Procedure for reproducing color copy by separating the color image into magenta, cyan, yellow, and black. The resulting four-color plates, when printed in register, reproduce the effects of color in the original image.

full color Color produced by the four-color process.

frequency polygon See **polygon–frequency**

furnish The mixture of fiber and other materials that is blended in the water suspension, or slurry, from which paper on board is made.

gang shoot To photograph several pages of original copy, such as pages of a book or booklet, at the same time. The pages are arranged on a surface so that after printing, folding, binding, and trimming they will be in numerical sequence. Also, the halftone photographing of illustrations arranged in a composite to be reproduced as a single plate.

gather To collate signatures in the order in which they are to appear in a book or journal.

gesso Plaster-like material made of chalk and white pigment, used to prepare a surface for painting.

ghosting Depiction of the central object of an image with full tonal values and the area around the object in flat or less pronounced tones. The heavier tones emphasize the central object; also called **fadeback**.

glossy prints Photographic prints on a shiny-finish paper. Prints intended for reproduction are usually made on such paper.

goldenrod A specially coated masking paper of yellow or orange color, used to assemble and position negatives for exposure onto press plates.

graph A diagram that shows amounts, frequencies, trends, or relationships of data. Examples include line graphs, bar graphs, histograms, pie graphs, etc.

graph–line A figure drawn on a coordinate grid expressing the continuous relationship of two or more variables; trends are depicted.

graph–semi-logarithmic line A figure drawn on a coordinate grid with one logarithmic axis expressing the continuous relationship of two or more variables; proportional and percentage relationships are shown.

graphic art 1. Art represented by drawing or imposing on a flat surface an image that communicates a message. 2. The methods, processes, and techniques employed in these arts.

gravure A printing process involving the transfer of ink from sunken images engraved on a plate cylinder.

gray scale A graded strip of gray tones, ranging from white to black, placed at the side of original copy during photography, or beside the negative or positive during plate exposure, to measure the tonal range obtained.

ground materials Surfaces on which the illustrator draws.

gutter margins Inner margins of two facing pages in a book, journal, or other publication.

halftone The reproduction of a continous tone original, such as a photograph, in which detail and tone values are represented by a series of evenly spaced dots of various sizes and shapes. The dot areas vary in direct proportion to the intensity of the tones they represent.

halftone screen A patterned screen (usually a dot pattern) placed between the continuous tone original copy and the photographic film in a process camera to break up the continuous-tone image into a photographic image composed of closely spaced dots. The densities of the resulting dots correspond to the tonal densities in the original copy.

hard copy A word processing, data processing, or typesetting term encompassing any output from a machine, or as a result of machine processing, which is readable copy on paper or film.

hard covers Covers consisting of rigid or flexible boards covered with cloth, paper or other material, used in the binding of books. See **case bound**.

highlight The lightest or whitest parts in a photograph, represented in a halftone by the smallest dots or the absence of dots.

histogram A graph on a coordinate grid used to visualize the distribution of continuous, grouped data. Frequency is shown on the Y axis, classes are on the X axis.

illustration board Laminated "sandwich" of high-quality drawing paper on one or both sides of a thick central board, available as cold press (rough) or hot press (smooth).

imposition The laying out of pages in a press form so that they will be in correct sequence after the printed sheet is folded.

intensities The extreme strengths, degrees, or amounts of ink.

internegative The negative from which a color print or transparency will be made.

keyline An outline drawn on illustrations to indicate where tone or color elements are to be printed.

key plate The plate, containing the most detail, used as a guide for the register of other colors in the four color process.

leaders Lines pointing to elements of an illustration. Can also mean rows of dashes or dots used to guide the eye across a page, as in a table of contents.

leading The amount of space between the lines of type, measured from baseline to baseline; always expressed in points. For example, 8/9 point size of type is 8, amount of leading is 9 points. This may also be expressed as 8 point type, 1 point lead.

line copy Illustrations containing black lines or portions of a line on a white background or vice versa, with no shades of gray.

line drawing A drawing containing no grays or middle tones. In general, any drawing that can be reproduced without the use of halftone techniques.

lithography A planographic method of printing that utilizes the chemical repulsion between water and grease to separate the printing from the non-printing areas. The printing design is greasy and ink-receptive, while the balance of the plate is damped with water to make it ink-repellent.

makeup The arrangement of printing elements to compose a page or other printing forms; sometimes called a **layout**.

map–choropleth A map using gray tones to represent rates, ratios, or frequencies within political or geographic areas (for example, to display demographic data, such as rates of birth and morbidity, population density and change, etc.).

map–flowline A map using arrows of different direction, length, and width to display the flow and amount of data in a geographic area (for example, flow of traffic, waves of migration, etc.).

map–isoline A map using lines to join points of equal value to indicate boundaries of various quantities (for example, temperature, rainfall, altitudes, etc.).

map–proportional A map that displays rates, ratios, or frequencies within political areas by distorting the area in proportion to the numerical value of the data.

matte finish Dull finish without gloss or luster.

mechanical 1. Camera copy showing exact placement of every element, and carrying actual or simulated type and artwork. 2. Page or layout prepared as an original for photomechanical reproduction; see **paste-up**.

middle tone The tonal range between highlights and shadows of a halftone.

moiré Undesirable patterns occurring when reproductions are made from halftones, caused by conflict between the ruling of the halftone screen and the dots or lines of the original; usually due to incorrect screen angles.

Mylar A polyester film, made by DuPont, specially suited for stripping because of its mechanical strength and dimensional stability.

negative Photographic image on film, in which black values in the original subject are transparent and white values are opaque;

light grays are dark and dark grays are light. The print prepared from a negative is a positive and shows the same tonal values as the original image that was photographed on film.

nib The business end of a pen.

offprints Additional sheets printed with the initial print order, which are cut into single pages and assembled as separate articles, bound by stapling. Offprints are commonly used by authors of journal articles. See also **reprints**.

offset press Rotary press in which the inked image on the pressplate is first printed onto a rubber blanket; the ink impression on the blanket is then offset onto the sheet of paper. Also known as **offset lithography**.

opacity The property of paper which minimizes the "show-through" of printing from the back side or the next sheet. The ability to impede light from passing through is a measure of the paper's opacity.

optical character recognition (OCR) System of employing an optical device which scans an image and "reads" the characters for phototypesetting.

ordinate The vertical or Y axis of a two dimensional coordinate grid or any other line parallel to the vertical axis.

original art The artwork, mechanical, or other material furnished for printing reproduction; usually refers to photographs or drawings to be processed for printing.

overlay Sheet of film or paper that contains additional graphics to be printed with a drawing. An overlay may be attached or hinged to original art for inscribing instructions to the printer.

Pantone Matching System (PMS; trademark of Pantone, Inc.) Coordinated system for matching printing inks and coordinating color reproduction and color reproduction materials.

paste-up The assembly of type elements, illustrations, etc., into final page form, ready for photographing.

perfect binding A binding method, usually for paper-covered books, with adhesive (glue) the only binding medium; also known as adhesive binding.

photocomposition See **phototypesetting**

photocopy All of the camera-ready or typeset copy used for an order; includes text, artwork, halftones, etc.; also called camera copy or original.

Photomechanical Transfer (PMT) Camera-generated positive print used for paste-up and for making paper contacts without the need for a negative.

phototypesetting The process of setting type by projection of type images onto photosensitive film or paper. The resulting positive prints are used to prepare mechanicals for printing plates.

pica A printer's standard measurement for length of lines and depth of type pages. One pica equals approximately 1/6 inch or 12 points.

pictogram A variation on the bar chart using a series of identifying symbols to represent the data.

plate A printed illustration. Also, a composite of grouped subjects.

plate–photographic A glass or film base coated with a photographic emulsion.

plate–printing A thin metal, plastic, or paper sheet that carries the printing design used in offset lithography. Also known as an offset plate.

ply A layer of paper or paperboard, several of which may be laminated together to form plies.

PMS See **Pantone Matching System**

PMT See **Photomechanical Transfer**

point A printer's unit of measurement, used principally for designating type sizes. There are 12 points to a pica; approximately 72 points to an inch.

polygon–frequency A graph to illustrate frequency distribution. The area enclosed in the polygon represents the distribution of continuous, grouped data.

press proofs Actual press sheets to show image, tone value, and color. A few sheets are run and approval received from the customer before printing the job.

press run Total number of copies of a publication printed during one printing.

printer's thumb The margin around a drawing; used for handling to prevent damage to the artwork.

process color The four color process reproduction of the full range of colors by use of four color plates (magenta, cyan, yellow, and black).

projection Various methods for depicting objects in three dimensional form.

projection–axonometric Drawing that shows the inclined position of an object with respect to the planes of projection. There are three types of axonometric projection: isometric, dimetric, and trimetric.

projection–dimetric Projection characterized by two faces inclined at equal angles.

projection–isometric Projection characterized by three faces inclined at equal angles.

projection–oblique Projection characterized by one face coincident with the picture plane. Most useful technique for charts.

projection–perspective (one point) Representation of an object so that it appears to have natural dimensions by drawing projection lines receding to one point.

projection–perspective (two point) Representation of an object so that it appears to have natural dimensions by drawing projection lines receding to two points.

projection–trimetric The representation of an object which has three unequal axes intersecting at right angles.

pyramid–age and sex or population Paired frequency distributions, depicting population by age and sex.

ream A standard parcel of paper, varying from 480 to 516 sheets depending on thickness and weight of the paper.

reflection copy Original copy for reproduction that is on an opaque material and must be photographed by light reflected from its surface. Examples are photographs, paintings, and dye-transfer prints (see **transmission copy**).

register 1. In printing, to align a type of page so that it exactly backs the type page on the reverse side of the sheet. 2. To match the position of successive color impressions. 3. The alignment of any corresponding elements of an image or impression.

register mark A symbol—⊕—that is placed on at least three sides of a drawing so that any overlay can be matched (registered) with it by having three or more register marks in the overlay placed precisely on top of the ones on the drawing. Three marks prevent the possibility of reversal which might occur with symmetrically placed marks. Preprinted register marks are available in cut-out or transfer form. They can also be hand-drawn.

reprints Articles originally printed in a book or journal for which negatives are restripped and reprinted. The articles are produced in booklet form. Reprints are more expensive to produce than offprints.

resolution In computer graphics, the fineness of detail in a reproduced spatial pattern, expressed in points per inch or points per millimeter.

reverse Type appearing in white on a black or color background or in a dark area of a photograph.

Ross board Clay-coated drawing paper with fine stipple surface (no longer manufactured).

routing Removal of unwanted parts of a printing plate. Usually performed with a graver or gouge.

rubber blanket Rubber sheet covering one cylinder of an offset press. The blanket receives the inked impression from the printing plate and transfers it to sheet-fed paper.

Rubylith Red light-safe masking and stripping film coated on a polyester backing sheet.

saddle wire stitching A binding method that inserts sections into sections and fastens them with wires (stitches) through the middle fold of the sheets; also called **saddlestitch**.

scattergram A graph showing the presence or absence of relationship between two variables.

scratchboard Laminated layers of paper coated with a mixture of white clay and binders. Usually white but sometimes inked to provide a solid black surface. Used by artists for ink work, especially where white texture on a black background is desired. The white drawing is scratched out of the black area inked by the artist or the black scratchboard.

screen patterns (shading medium, line tints) Films printed with screens or textures, mounted on a translucent backing, which can be peeled or rubbed off onto an illustration.

serif Finishing stroke or line projecting from the end of the main stroke of a letter.

shadow The darkest parts in a drawing or photograph.

sheet fed Designates a printing press to which paper is fed in sheets rather than in rolls, or webs.

show-through Visibility of printed matter on the opposite side of a sheet. The effect is due largely to poor choice of paper, but it may also be caused by excessive ink penetration of the paper or too heavy an impression between the printing plate and the impression cylinder; also called **see-through**.

shrink wrap A method of packaging in plastic film. The material to be wrapped is inserted into a folded roll of polyethylene film, which is heat-sealed around it. The package then goes through a heat tunnel where the film shrinks tightly around the package; also called **plastic wrap**, **plastic shrink wrap**.

signature A folded, printed sheet, forming one section of a book or a journal.

silhouetting Opaquing out the background around a subject on a halftone negative to produce a silhouette of the subject.

sizing The treatment of paper which gives it resistance to the penetration of liquids (particularly water) or vapors.

soft covers Another term for paperback or paperbound books. Also known as **perfect binding**.

specifications Detailed description of all required parts of a job. The "specs."

spine The back of a bound book; also called the backbone.

spot color Flat area of color in a printed piece; often added to black and white artwork for emphasis.

stereogram A diagram that represents objects in an apparently three dimensional form with height, width, and depth representing X, Y, and Z axes.

stippling Dots, either round or linear (but ideally round), repeated in an area so as to produce a pattern or the impression of tone.

stock Paper, paperboard, or other paper product on which an image is printed, copied, or duplicated.

stripping 1. The act of positioning or inserting copy elements in negative or positive film to make a complete negative; the positioning of photographic negatives or positives on a lithographic flat or form. 2. The condition under which steel rollers fail to take up the ink on lithographic presses and instead are wet by fountain solution.

tack (of ink) The stickiness or adhesive quality of the ink.

tear sheet 1. Any page torn from a book, with corrections or changes marked on it. 2. Sheet extracted from a publication that contains an advertisement or other matter.

technical pens Pens with hollow-point inflexible nibs of various diameters, containing a moveable wire core. Ink flows from a reservoir through the nib to create a smooth line of specified weight.

tone Gray-appearing areas, either truly gray, as created by pencil or ink wash, or illusory, as created by lines and dots; also applies to color gradation.

tooling Handwork on an engraving or printing plate to improve its printing qualities. Commonly refers to the space or border that is stripped in between margins of group-arranged illustrations.

transfer Sheet or gelatin-like film containing the image that is to be transferred to a metal printing surface.

transfer type Letters, numbers, and symbols on clear sheets, which can be applied to many surfaces by rubbing or burnishing after positioning.

transmission copy Original copy that allows light to pass through it, such as color transparencies (see **reflection copy**).

transparency A monochrome or full-color photographic positive or picture on a transparent support, the image intended for viewing and reproduction by transmitted light. Smaller transparencies mounted in 2 × 2 inch frames are often called **slides**.

trim size The finish size of a journal, book, or other printed piece after trimming.

two color Use of two colors in printing; usually black with one other color.

type face A style or design of type encompassing shape, weight, and proportions which make it distinct from other type faces.

uncoated paper Paper generally made on a conventional Fourdrinier machine and machine dried; may or may not be surface sized.

Van Dyke A photographic image made on inexpensive photopaper sensitized with a mixture of iron and silver salts; a brownprint. Frequently used by offset lithographers as a proof from a negative, to be sent to the customer as a means to check the layout or other corrections, for content only; also known as **silverprint**, **blue**, **brownline**.

variable A function with no fixed value. Its values may vary dependently or independently of other variables.

variable-dependent A function whose values are affected by changes in the other variables; usually plotted on the Y axis.

variable-independent A function whose values are unchanged by changes in other variables, for example, time; usually plotted on the X axis.

vellum 1. High-quality translucent or transparent tracing paper. 2. Opaque and antiqued high-quality paper with a "kid" finish. 3. Originally, a writing surface made from calfskin.

wash Application of watercolor, usually in large areas of an illustration. Also refers to technique used for monochromatic continuous-tone medical illustrations.

web press A printing press to which paper is fed in rolls or webs rather than in sheets.

weight (paper) Designated fixed weight of 500 sheets (one ream) of paper of the basic sheet size, used as the basis for measuring the substance of paper by weight. Different classes of paper, such as writing, cover, and book, have fixed measuring sizes which determine the designated weights. The weight of 1,000 sheets of paper is called the "M" weight, found by doubling the 500-sheet weight; also called **basis weight**.

window An open or clear area that permits light to pass through; usually large area on a negative or hand-cut opening on a masking sheet to expose the image.

word spacing Adjustment of spacing between words, particularly to shorten or extend a line in order to achieve justification (alignment) at the margins of a printed line.

wove paper A paper finish that has no visible laid lines.

xerography A copying process that utilizes electrostatic forces to form an image.

Zip-a-Tone Trade name for a series of screen patterns imprinted on plastic sheets that can be used to achieve tone of various kinds on art work.

CHAPTER XIII
ANNOTATED BIBLIOGRAPHY

Bibliography Contents

Guides for Authors and Editors

Allen A: Steps toward better scientific illustration. Lawrence, KS, Allen Press, 2 ed., 1977.

Guide to preparation and production of line and halftone illustrations with consideration of special problems of foldout illustrations, micropublications, 300-line halftone screens, mathematical equations, and chemical formulas.

American Psychological Association: Publication manual of the American Psychological Association. Washington, DC, American Psychological Association, 3 ed., 1983.

The sections on figures and graphs are useful to authors and could serve as models for editors in shaping their own guidelines (pp. 94–105).

CBE Style Manual Committee: Council of Biology Editors style manual: A guide for authors, editors, and publishers in the biological sciences. Bethesda, MD, Council of Biology Editors, 5 ed., 1983.

Concise, general guide for authors and editors covering all aspects of journal publishing. Chapter on preparation of tables is outstanding.

Fishbein M: Medical writing: the technic and the art. Springfield, IL, Charles C Thomas, 4 ed., 1978.

Chapter 13 on illustrations (pp. 149–164) summarizes information on choosing and preparing illustrations and provides examples that distinguish the adequate from the inadequate. A summary of rules for authors is pertinent.

Reynolds L, Simmonds D: Presentation of data in science. The Hague, Martinus Nijhoff, 1981.

The subtitle (Publications, slides, posters, overhead projections, tape-slides, television. Principles and practices for authors and teachers) indicates the scope. A clear exposition of the subject from manuscript to the printed product. Coverage of halftone and color reproduction is sparse whereas emphasis is given to tables, graphs, slides, posters, and television.

Skillen ME, Gay RM: Words into type. Englewood Cliffs, NJ, Prentice-Hall, 3 ed., 1974.

Another classic for authors and editors. Sections on typographical style (pp. 243–337), mathematics (pp. 283–288), and illustration (pp. 491–534) are helpful for orientation.

University of Chicago Press: A Manual of Style. Chicago, IL, University of Chicago Press, 13 ed., 1982.

Standard reference work with concise coverage of all aspects of thesis, periodical, and book publication. This new edition has expanded coverage of illustrations and tables.

Illustration Techniques

Cardamone T: Advertising agency and studio skills. New York, NY, Watson-Guptill Publications, 1981.

Practical guide to preparation of art for reproduction. Covers materials, methods, terminology, equipment, and problem solving.

Gray B: Studio tips for artists and graphic designers. New York, Van Nostrand Reinhold, 1976.

Illustrated "how-to" guide showing practical methods of preparing and handling artwork.

Hodges ERS (ed.): The Guild handbook of scientific illustration. (in preparation by the Guild of Natural Science Illustrators; publication planned for 1988).

An encyclopedic collection of contributions by more than 50 authors. Includes general principles of scientific illustration, rendering techniques, care, measurement, and manipulation of animal and plant specimens, technological tools (microscopy, photography computers), charting data, reproduction and printing, and business considerations.

Holmgren NH, Angell B: Botanical illustration: preparation for publication. Bronx, NY, New York Botanical Garden, 1986.

Two prominent artists collaborated to produce this remarkably concise guide for the preparation of publication quality botanical illustrations. The emphasis is on techniques for producing line art (tools and materials, sizing, and labeling). A brief section on the use of maps in botany is included.

Ridgeway JL: Scientific illustration. Stanford, CA, Stanford University Press, 1979.

First published in 1938 and reprinted unrevised, this work is "primarily intended to aid students and others engaged in the preparation of manuscripts that require illustrations." Addresses considerations such as posing specimens (orientation) and direction of lighting in photographs.

Sayner DB, Menhennet GB: Drawing for scientific illustration. Tucson, AZ, University of Arizona Press, 1969.

Each of 4 booklets covers a different topic: Technique and Rendering, Production, Professional Aids and Devices, Maps-Graphs-Diagrams. Booklets demonstrate concepts rather than specific details. Many helpful hints for illustrators, especially for geological illustration, Mr. Sayner's specialty.

Wood P: Scientific illustration. A guide to biological, zoological, and medical rendering techniques, design, printing, and display. New York, NY, Van Nostrand Reinhold, 1979.

The title explains it all. Illustrated guide to planning, producing, and converting the illustration to the printed page. One of the few publications with a section on preparing graphics for exhibits (poster sessions, etc.).

Zweifel FW: A handbook of biological illustration. Chicago, IL, University of Chicago Press, 1961.
A brief guide for students, authors, editors, and illustrators on the preparation of various kinds of illustrations for publication. This early classic is being revised by the author for a new edition due to appear in 1988.

Photography Techniques

Blaker AA: Handbook for scientific photography. San Francisco, CA, W. H. Freeman, 1977.
General guide to laboratory photography, including photomicrography.

Eastman Kodak Company: Basic photography for the graphic arts (Publication Q-1). Rochester, NY, Eastman Kodak Co., Graphics Market Division, 3 ed., 1977.
Basic explanation of photographic procedures used in preparation of halftone negatives and prints.

Engel CE (ed.): Photography for the scientist. New York, NY, Academic Press, 1968.
A general reference work by 14 authors, mostly British and German. There is some duplication, but a good range of viewpoints is represented. Needs updating.

Hansell P (ed.): A guide to medical photography. Baltimore, MD, University Park Press, 1979.
All aspects of clinical photography are covered. Informative chapters on reproduction of graphic originals and on ethical considerations of clinical photography are included. Most contributors are distinguished European photographic illustrators.

Loveland RP: Photomicrography: a comprehensive treatise. New York, NY, John Wiley & Sons, 1970.
Two volume treatise on the subject, including optics, illumination, photographic principles, and processing. Needs updating but remains one of the best works of its kind.

Halftone Reproduction

Du Pont de Nemours & Co.: The contact screen story. Wilmington, DE, E. I. Du Pont de Nemours & Co., Photo Products Department, 1972.
Clear, simplified explanation of the halftone process, with helpful tips on camera technique.

Eastman Kodak Company: Halftone methods for the graphic arts (Publication Q-3). Rochester, NY, Eastman Kodak Co., 3 ed., 1976.
Basic and advanced techniques used in the production of halftones are succinctly described and illustrated. Especially clear descriptions of the halftone dot and its relation to the gray scale.

Eastman Kodak Company: Kodak contact screens (Publication Q-21). Rochester, NY, Eastman Kodak Co., 1975.
Brief description of halftone screens and their uses.

SD Warren Co.: Black halftone prints (Bulletin No. 3). Boston, MA, S. D. Warren Co., 1980.
An atlas in which several photographs are reproduced with various special-effect halftone screens on different paper stocks. Illustrated criteria for judging good photographic copy are especially useful.

Sternbach H: Halftone reproduction guide. Great Neck, NY, Halftone Reproduction Guide, 1965.

An unusual 100 page atlas that displays a single subject reproduced with single dot, duotone, and triple dot halftone screens combined with variations in gray scale, color inks, and paper stock. The effects of these variables on the same subject throughout the book are immediately apparent and very instructive.

Color Reproduction

Anon.: Selectone four color process guide. 2 ed., (publisher and date not given).

This spiral bound 87 page album displays 4 colors in combinations of 10% increments for each color (a total of 9,410 hues) using 133 line screens. By matching swatches of graded color percentage in grid-like arrangement, the resulting hue can be predicted.

Kelly KL, Judd DB: Color universal language and dictionary of names (National Bureau of Standards Special Publication 440). Washington, DC, U.S. Department of Commerce, 1976.

The American standard for color identification. This somewhat complicated "dictionary" brings together all the established color-order systems and methods of designating color. Color names are accompanied by samples of their corresponding color hues.

Eastman Kodak: Basic color for the graphic arts (Publication Q-7). Rochester, NY, Eastman Kodak Co., 1977.

Descriptions of color principles, direct and indirect color screen methods, and color reproduction equipment are enhanced by easily understood diagrams.

Pantone, Inc.: Pantone Color Formula Guide. Moonachie, NJ, Pantone, Inc., 18 ed., 1985.

This is a fold-out series of 8¼ × 2 inch strips of color hues identified by the Pantone system of catalog numbering. Each sample specifies the proportional color components used to achieve each hue. Useful for specifying exact color requirements for color production.

Tables, Graphs and Maps

Batschelet E: Introduction to mathematics for life scientists. Berlin, Heidelberg, New York, Springer Verlag, 3 ed., 1979.

Chapter 3 describes relations and functions, including line graphs. Chapter 7 describes graphical methods, including clear mathematical explanations of nonlinear scales, semilogarithmic plots, double-logarithmic plots, triangular charts, nomography, and pictorial views (three-dimensional renderings). Chapter 13 describes probability, including continuous distributions (frequency distributions).

Brouwer O: The cartographer's role and requirements. Scholarly Publishing, 14(3): 231–242, 1983.

An article delineating the roles of the editor, author, and cartographer in the production of well-designed maps.

Clark N: Tables and graphs as a form of exposition. Scholarly Publishing, 19(1): 24–42, 1987.

The author proposes that tables and graphs should be designed (and analyzed) by a set of descriptors that include the source of data, identification of the observors, the matter or subject reported, the function (nature of the event presented), the space (location of the event displayed), the time of the event, and the domain or nature of the values assigned. This article makes a convincing case for the author's conclusion that the basic rules for verbal and graphic presentation are the same (or *should* be).

Cleveland WS, McGill R: Graphical perception: theory, experimentation, and application to the development of graphical methods. Journal of the American Statistical Association, 79: 531–554, 1984.

The authors propose a theory for the construction of graphic displays based on elements of graphic perception. This article reports the authors' experiments in perception and their results are applied to a variety of graph forms with improvements that should enhance the reader's comprehension.

Cleveland WS: The elements of graphing data. Monterey, CA, Wadsworth Advanced Books and Software, 1985.

A thoughtful book on the construction of graphs, based on the studies of data perception performed by Cleveland and his colleagues at the Bell Telephone Laboratories. Chapter 2 on the principles of graph construction is especially valuable.

Herdeg W (ed.): Graphis diagrams. Zurich, The Graphis Press, 1976.

A collection of graphs and diagrams to illustrate effective graphic visualization of abstract data.

Holmes N: Designer's guide to creating charts and diagrams. New York, NY, Watson-Guptill, 1984.

Widely known for his innovative approach to "illustrated" graphs by readers of Time magazine, Holmes' designs make statistics palatable for the common reader. Although unconventional for presentation of "scientific" data in conventional "scholarly" journals, the visual impact of the examples in this book deserves study.

Lawrence GRP: Cartographic methods. London, New York, NY, Methuen, 2 ed., 1979.

Materials, techniques, and equipment required for topographic and thematic mapping.

Lockwood A: Diagrams. A visual survey of graphs, maps, charts and diagrams for the graphic designer. London, England, Studio Vista, 1969.

A text, fully supported by many illustrations, dealing with statistical and explanatory diagrams and maps. An excellent source book for ideas.

MacDonald-Ross M, Waller R: The transformer. Penrose Annual. Vol. 69, 1976, pp. 141–152.

An examination of the range of problems involved in putting the experts' message in a form the ordinary person can best understand and use. Uses as examples a table, an isotype graph, and an algorithm.

MacDonald-Ross M, Smith E: Graphics in text: a bibliography. London, Milton Keynes: Institute of Educational Technology, The Open University; 1977.

A bibliography of literature relating to graphic presentation.

MacGregor AJ: Graphics simplified: how to plan and prepare effective charts, graphs, illustrations, and other visual aids. Toronto, University of Toronto Press, 1979.

A concise no-nonsense guide to the preparation of charts and graphs, selection of the most appropriate illustrations, and planning of the best graphic approach to the data problem. The clear exposition of illustration principles makes this work especially suitable for authors and editors.

MacGregor AJ: Graphing the frequency distribution. The Journal of Audiovisual Media in Medicine, 3: 17–20, 1980.

A clear guide to graphing frequency distributions.

Murgio MP: Communications graphics. New York, Van Nostrand Reinhold Co., 1969.

A search for this out-of-print monograph is worth the effort. Covers the why and how of chart making: when to chart, what to chart. Excellent discussion of the variables to be considered in slide making and viewing.

Robinson AH, Sale RD, Morrison J: Elements of cartography. New York, NY, John Wiley, 4 ed., 1978.

A classic textbook on map-making.

Schmid CF, Schmid SE: Handbook of graphic presentation. New York, NY, John Wiley & Sons, 2 ed., 1979.

A comprehensive, well illustrated handbook which discusses basic principles and techniques of graph design, several common types of graphs, maps, projection techniques, and the role of the computer in graphic presentation.

Simmonds D: Medical chartist's dilemma. Medical and Biological Illustration. 26: 153–158, 1976.

An attempt to enumerate standards for artists and chartists.

Simmonds D, Gragg G: Charts and graphs. Guidelines for the visual presentation of statistical data in the life sciences. Lancaster, England, MTP Press, Ltd., 1980.

Compiled by members of the Institute of Medical and Biological Illustration, the book is intended as a guide for preparation of charts and graphs for publication or of teaching slides, television artwork, and tape/slide programs. Appendices cover abbreviations, flow chart symbols, materials and equipment, and other useful information.

Spear ME: Practical charting techniques. New York, NY, McGraw-Hill Book Company, 1969.

A handbook dealing with preparing graphic presentations of statistics or economics. The book discusses development of illustrations for displays, meetings, reports, and books, and the techniques for their preparation.

Truran HC: A practical guide to statistical maps and diagrams. London, Heinemann Educational Books, 1975.

A short (60 pages) useful guide to statistical diagrams, their construction and merits.

Tufte ER: The visual display of quantitative information. Cheshire, CT, Graphics Press, 1983.

An exceptional presentation of theory and design of data graphics. Acceptable and unacceptable solutions to the graphic presentation of statistical problems are clearly illustrated.

US Department of Health and Human Services: Descriptive statistics. Tables, graphs, & charts. Atlanta, GA, Center for Disease Control, 1977 (Publication no. 00-1834).

Concise presentation of the principles for construction of tables and the most frequently used graphs and charts.

Computer Graphics

ACM transactions on graphics. Baltimore, MD, Association for Computing Machinery, Inc.

A quarterly journal that emphasizes innovative applications of computer graphics, graphics systems design, image synthesis, geometric modeling, and algorithm design.

Crovello TJ: Computers in the life sciences. IV: Computer graphics. Bioscience, 31: 697; 1981.

A brief introduction to the use of computer graphics for the biologist.

Freeman H: Interactive computer graphics. New York, NY, The Institute of Electrical and Electronic Engineers, 1980.

Papers presented at an annual conference on computer graphics with current bibliography.

George JE, Vinberg A: The display of engineering and scientific data. IEEE Transactions on Professional Communications, 25: 95–97, 1982.

An intriguing article which deals with the display of progressively more selective and communicative graphic information. One problem is handled eight ways, all by computer.

Graedel TE, McGill R: Graphical presentation of results from scientific computer models. Science, 215: 1191–1198, 1982.

Review of computer graphics as applied to displays of concentration, rates, multiple related variables, and comparison of results with data.

Hyman A: The computer in design. London, Studio Vista, 1973.

Introduction to computer graphics, including three-dimensional graphics, and principles of design. Includes a still useful glossary and bibliography of earlier publications.

Kerlow IV, Rosebush J: Computer graphics for designers and artists. New York, NY, Van Nostrand Reinhold, 1986.

Designed for illustrators with no previous computer experience, this well-designed, clearly illustrated manual takes the reader through basic computer concepts, hardware, software, and peripherals to the applications for two- and three-dimensional graphics production. Perhaps the best single source for authors, illustrators, editors, and publishers.

Marcus A: Computer-assisted chart making from the graphic designer's perspective. Computer Graphics, 14: 247–253, 1980.

Awareness of graphic design is necessary to prepare effective charts. This article discusses basic design principles as they apply to computer-produced charts.

Pollack V: Becoming comfortable with computer graphics. San Diego, CA, Hewlett-Packard, 1980.

Includes definition, history, discussion of types of graphs, graphic concepts, drawing axes and grids, plotting data and shapes, labeling, characteristics of different graphic systems, programming languages, and a useful glossary.

SIGGRAPH: Computer graphics. New York, Association for Computing Machinery, Inc.

Computer Graphics is a quarterly journal of the Special Interest Group on Graphics (SIGGRAPH). Articles cover all aspects of graphical man/machine communication: hardware, language and data structures, methodology, and applications.

Smith Jr. HR (ed.): Computergraphics in biocommunication. Journal of Biocommunication (special computergraphics issue) 11: (no. 2), May, 1984.

Eleven contributions from authors with practical experience in all aspects of computer graphics, including equipment, financing, medical arts management, illustration standards, and networking.

van Dam, A: Computer software for graphics. Scientific American, 251 (no. 3): 146–159, 1984.

A clear, basic description of the principles involved in producing illustrations by computer. Emphasis is on line and tone art.

Whitted T: Some recent advances in computer graphics. Science, 215: 767–774, 1982.

Informative review of current trends in design and various applications of computer graphics.

Camera-ready Copy, Desktop Publishing, and Graphic Arts Production

American Society of Biological Chemists: The Journal of Biological Chemistry, Instructions to Authors, Miniprint Section. Bethesda, Maryland, American Society of Biological Chemists, Vol. 262, No. 1, 1987.

Requirements for the use and directions for the preparation of miniprint. Intended as an aid to the typist.

Ballinger RA: Art and reproduction: graphic reproduction techniques. New York, NY, Van Nostrand Reinhold, 1977.

Introductory guide to the production of illustrated material. Many helpful tips on "how to do it."

Craig J: Production for the graphic designer. New York, Watson-Guptill Publications, 1974.

Brief but succinct coverage of typesetting printing, papers, inks, folding, binding, and preparation of mechanicals.

Field JN (ed.): Graphic arts manual. New York, NY, Arno Press, 1980.

A comprehensive guide to creating, producing, and purchasing printed material. The manual is divided into 16 chapters comprising 152 sections, written by 94 contributors. Topics range from design and manuscript preparation, through the usual stages of graphic production, to discussions of trade practices (printer-customer relations, copyrights, patents, trademarks, and trade customs).

IFSEA, CIBA, ELSE: Model guidelines for the preparation of camera-ready typescripts by authors & typists. London, Ciba Foundation, 1980.

Suggestions and guidelines from an International Federation of Scientific Editors Associations-Ciba Foundation Workshop held May 18, 1979. This 47-page pamphlet is available at a nominal fee from the editorial department of the Ciba Foundation, London.

Kleper ML: The illustrated handbook of desktop publishing and typesetting. Blue Ridge Summit, PA, Tab Books Inc., 1987.

At present, most monographs on desktop publishing are linked to particular configurations of hardware and software. This book is remarkably comprehensive in its descriptions of alternative possibilities for text creation tools, telecommunication methods, typesetting hardware, and output devices. Clear descriptions of publishing applications are provided. The annotated bibliography is especially valuable.

Latimer HC: Preparing art and camera copy for printing. New York, NY, McGraw-Hill, 1977.

A detailed guide on the preparation of materials for photomechanical processes. Although commercial applications are used as examples, the process descriptions are unusually clear and applicable to scientific publication, for example, designing for standard paper sizes, specifying type size, copy fitting, cropping and scaling of illustrations, etc.

Lem DP: Graphics master 2. Los Angeles, CA, Dean Lem Associates, 2 ed., 1977.

A spiral-bound compendium of data on graphic arts production. Contains abbreviated information (usually in tabular form) on halftones, color work, paper, inks, presses, type faces, copy fitting, binding, etc.

Stevenson GA: Graphic arts encyclopedia. New York, NY, McGraw-Hill, 2 ed., 1979.
A comprehensive reference work on technical aspects of graphic arts production, arranged in dictionary form. Technical subjects are lucidly explained and carefully illustrated. Descriptions of production equipment (typesetters, presses, etc.) are especially helpful.

Stone B, Eckstein A: Preparing art for printing. New York, Van Nostrand Reinhold, 1979.
Clear descriptions and illustrations of the printing and plate-making processes and the preparation of mechanicals.

The Baxter Group: Macintosh desktop design. Sunnyvale, CA, The Baxter Group, 1984.
Part of a series of publications on desktop publishing using Apple Computer's Macintosh system, this publication includes descriptions of principles, design, and page composition.

Ulick T: Personal publishing with PC PageMaker. Indianapolis, IN, Howard W. Sams & Co., 1987.
The use of Aldus Corporation's software program, *PC PageMaker*, is described, with practical hints on selection of hardware and software, page design, operational procedures, and lists of suppliers.

The Williams & Wilkins Company: Preparation of camera-ready manuscripts. Baltimore, Williams & Wilkins, 1978.
Guidelines for the preparation of camera-ready material for book production.

Publishing and Printing

Blair R: The lithographers manual. Pittsburgh, PA, Graphic Arts Technical Foundation, 8 ed., 1988.
Probably the most comprehensive reference work on lithography and related subjects. Extensive treatment of camera work, color separation, halftone production, photocomposing, plate making, proofing, inking, paper specifications and binding.

Craig J: Designing with type. New York, NY, Watson-Guptill Publications, 1971.
Introduction to the fundamentals of typography. Basic information on letter and word spacing and the consideration of alphabets as design elements.

Craig J: Phototypesetting. A design manual. New York, NY, Watson-Guptill Publications, 1978.
Exceptionally lucid presentation of terminology, design features, copyfitting, and phototypesetting systems.

Edwards Brothers: Graphics arts glossary. Ann Arbor, MI, Edwards Brothers, 1978.
A valuable glossary for the graphic arts with lists of terms for composition, camera work, plate preparation, press work, bindery, shipping, web press, and data processing.

International Paper Company: Pocket Pal: a graphic arts production handbook. New York, NY, International Paper Company, 13 ed., 1983.
A pocket-sized reference work which covers the entire process of printing. Contains the minimal amount of information needed for the editor's assistants.

Munday PJ, Radford HT: Illustration preparation and reproduction techniques. The publishers' needs. Medical and Biological Illustration, 26: 111–114, 1976.
Cautionary notes from the publisher to authors and editors. The message is not new, but the viewpoint is of interest because the publisher rarely speaks out in print. Useful reading for editors.

Rehe RF: Typography: how to make it most legible. Carmel, IN, 3 ed., 1979.
An unusual 80-page paperback treatise on the variables that influence perception of the printed word (comparative legibility of type fonts, type size, line width, leading, justified typography, etc.). The implications for design should be considered by illustrators, editors, and publishers.

Wijnekus FJM: Elsevier's dictionary of the printing and allied industries. Amsterdam, Elsevier Publishing Co., 1967.
A 583-page multilingual (English, French, German, and Dutch) technical dictionary with appendices and bibliography. A unique reference work that should be updated.

Book Design and Bookmaking

Association of American University Presses: One book/five ways. The publishing procedures of five university presses. Los Altos, CA, Association of American University Presses, 1978.
A "make-believe" book is analyzed by five university presses. Details of their acquisitions procedures, production and design, and sales and promotion are reported. An experiment which provides more information on the publishing process than any textbook to date.

Lee M: Bookmaking: the illustrated guide to design and production. New York, NY, R. R. Bowker, 2 ed., 1979.
The standard reference on the topic, in itself a model of design which emphasizes the problems of illustration and how to solve them.

Peters, J (ed.): The bookman's glossary. New York, NY, R. R. Bowker, 5 ed., 1975.
Definitions of technical terms relating to editing, printing, and marketing of books.

Legal Considerations

Crawford TC: Legal guide for the visual artist. New York, NY, Madison Square Press, 2 ed., 1986.
Chapters on rights of the artist and the copyright law are relevant to the working relationship between illustrators and the author-editor-publisher complex.

Crawford TC: The visual artist's guide to the new copyright law. New York, NY, Graphic Artists Guild, 1978.
Plain language description of the revised copyright law as it applies to graphic arts. The section on works for hire should be read by authors, editors, and publishers.

Graphic Artists Guild: Graphic artists guild handbook. Pricing & ethical guidelines. New York, NY, Graphic Artists Guild, 5 ed., 1984.
Guidelines for pricing and contracting illustrators' work. Chapters on medical and technical illustration are pertinent.

Latman A: The copyright law. Washington, DC, Bureau of National Affairs, 5 ed., 1979.
Basic reference work on the history, application, and requirements of copyright. Includes text of the 1909 and 1976 Copyright Acts.

Strong WS: The copyright book: a practical guide. Cambridge, MA, MIT Press, 1981.
Applications of the 1976 copyright law, concisely presented and discussed.

INDEX

Note: Page number in italics indicates a figure or table.

ILLUSTRATING SCIENCE: STANDARDS FOR PUBLICATION

DESIGNED BY DENNIS STILLWELL AND GEORGE LAWS

COMPOSED AT CAPITAL CITY PRESS, MONTPELIER, VERMONT
TYPESET IN BASKERVILLE

PRINTED AT ALLEN PRESS, LAWRENCE, KANSAS
ON 60# PRODUCTOLITH GLOSS AND 70# CENTURA GLOSS

BOUND BY ZAHRNDT'S, ROCHESTER, NEW YORK
IN HOLLISTON ROXITE VELLUM